Modern Earth Structures for Transport Engineering

Modern Earth Structures for Transport Engineering

Engineering and Sustainability Aspects

Ivan Vaníček, Daniel Jirásko and Martin Vaníček

CRC Press
Taylor & Francis Group
Boca Raton London New York Leiden

CRC Press is an imprint of the
Taylor & Francis Group, an **informa** business

A BALKEMA BOOK

CRC Press/Balkema is an imprint of the Taylor & Francis Group, an informa business

© 2020 Taylor & Francis Group, London, UK

Typeset by Apex CoVantage, LLC

Library of Congress Cataloging-in-Publication Data
Applied for

Published by: CRC Press/Balkema
Schipholweg 107C, 2316 XC Leiden, The Netherlands
e-mail: Pub.NL@taylorandfrancis.com
www.crcpress.com – www.taylorandfrancis.com

ISBN: 978-0-367-20834-9 (Hbk)
ISBN: 978-0-367-54603-8 (pbk)
ISBN: 978-0-429-26366-8 (eBook)
DOI: https://doi.org/10.1201/9780429263668

Contents

Preface

Earth structures together with foundation structures and underground structures fall under the more general subject matter of geotechnical structures. Under the term geotechnical structures earth structures are the oldest matter of concern, but rather paradoxically in comparison to foundation structures (foundation engineering) and to underground structures (tunnelling), the first monograph covering all aspects of earth structures was written only one decade ago (Vaníček and Vaníček, 2008). Briefly stated, we can use the following categorical explication for the division of geotechnical structures:

- Construction with ground for earth structures, where ground (soil) is the natural and most used construction material;
- Construction on the ground for foundation structures;
- Construction in the ground for underground structures.

This book will deal only with earth structures for transport engineering, although it is recognized that earth structures for water or environmental engineering also generate other interesting concerns. Embankments and cuts of motorways and railways comprise the main subject of the presented material. However, the main recommendations are valid also for airfields and parking areas.

Development in each profession is progressing in the form of individual steps, wherein a hierarchical part relates to new knowledge or to pressure from society to improve something, and a lateral part is more connected with the implementation of this knowledge into general practice. For earth structures, development was very slow up to the beginning of the 20th century, when design was mostly based on previous experience. The first significant advance is usually connected with the principle of effective stresses and the theory of consolidation (Terzaghi, 1925, 1943). Other advances were made in the 1950s to 1970s with new testing devices Bishop and Henkel (1957, 1962), the principle of stress path (Lambe and Whitman, 1969), and the implementation of numerical methods for the design of structures (Zienkiewicz, 1967).

Present-day demands on earth structures in transport engineering are strongly influenced by two aspects.

The technical engineering aspect involves structure safety, along with optimal design and implementation. From this perspective, all new knowledge should be applied. Nevertheless, the basic approach to safety and optimal design is strongly affected by geotechnical risk, to which this design and structure implementation is connected. However, to guarantee a certain level of safety for a certain structure, some standards should be accepted. For example, in Europe these standards are included in Eurocode 7 (EC 7) *Geotechnical Design* (EN 1997).

The second aspect is strongly connected with acceptance of the principle of sustainable development, which was specified at the highest level during the international "Environmental Summit" in Rio de Janeiro in 1992. Gro Harlem Brundtland, the Norwegian prime minister, used for example the following definition of sustainable development: "Development which responds to the needs of the present without compromising the capacity of future generations to respond to their needs." However, the sustainability approach strongly relates to two other principles, which are nowadays specified for earth structures in transport engineering – the principles of availability and affordability.

Both aspects form the essence of this book, and therefore the term "modern" can be used for this stage of development. The authors recognize with great appreciation support from the Czech Technical University in Prague where the research project CESTI (the Centre for Effective and Sustainable Transport Infrastructure) is being carried out.

Chapter 1

Introduction

Geotechnical engineering has always had a very close relationship with nature itself. This means that geotechnical engineering is not only about technical solutions but also about environmental considerations. However, an interest in and the problems associated with environmental protection are gradually causing the topic to acquire a special position within the wider branch of classical geotechnics.

A very significant step in this process was the state-of-the-art report "Environmental Geotechnics" presented by Sembenelli and Ueshita during the Xth International Conference SMFE in Stockholm 1981. Dating from 1994, the International Congress of Environmental Geotechnics has been organized by International Society for Soil Mechanics and Foundation Engineering (ISSMFE), later denoted as International Society for Soil Mechanics and Geotechnical Engineering (ISSMGE) – in Edmonton 1994, Osaka 1996, Lisbon 1998, Rio de Janeiro 2002, Cardiff 2006, New Delhi 2010, Melbourne 2014 and Hangzhou 2018.

The concept of sustainable development was accepted, as previously mentioned, in Rio de Janeiro during the international "Environmental Summit." Thereafter, this concept was gradually developed in various areas of human activities, including the construction sector (Vaníček, 2011). However, priority was given mostly to building engineering and especially to energy efficient buildings.

Any implementation within civil engineering has been more complicated, but nevertheless initial efforts are emerging, and this applies also for matters of geotechnical engineering (e.g. Vaníček and Vaníček, 2013a; O'Riordan, 2012; Vaníček et al., 2013; Correira et al., 2016; Basu et al., 2013a, 2013b; Correira, 2015).

The main aim of sustainable development is to provide economically competitive construction with higher utility value while also making lower energy demands, requiring lower raw material inputs and reducing the need for new plots of land. At the same time, the risk of danger to human health and life during natural disasters, accidents and unwanted events can be moderated.

In the field of transport infrastructure, some priorities have been defined in recent years, e.g. in the following contributions:

- Horizon 2020 Transport Advisory Group (TAG), May 2016;
- FEHRL Vision 2025 for Road Transport in Europe;
- ECTP reFINE (2012);
- ELGIP Position Paper (2016, 2018).

Basically, all emphasize three main aspects: sustainability, availability and affordability.

Sustainability involves emphasizing the increased resource efficiency of infrastructures. This comes through the development of more economical and environmentally acceptable earth structures. Therefore, attention is devoted to lowering energy use and the consumption of natural aggregates and attempting to save land.

Availability emphasizes increasing infrastructure capacity not only for current concerns but also for anticipated future changes, e.g. from the climate change perspective. Therefore the interaction of transport infrastructure with natural hazards such as in landslides, rock falls, or floods are studied very intensively – e.g. in the European project INTACT, and also by the authors themselves (Vaníček *et al.*, 2016; Jirásko and Vaníček, 2015; Jirásko *et al.*, 2017; Vaníček *et al.*, 2018). However, not only natural hazards but also accidents connected with human activities should be eliminated.

Affordability is strongly connected with a reduction of lifecycle costs. Therefore, new methods of checking structural conditions are being introduced to be able to predict deterioration in structures (ageing). Lifecycle cost also depends on demands for structure maintenance. Attention is thus focused on structures with low cost maintenance or on new, more efficient methods of structure maintenance.

The European Large Geotechnical Institutes Platform (ELGIP), in lecture "Geotechnical Risk Reduction for Transport Infrastructure" presented in Brussels for the European Council for Construction Research, Development and Innovation (ECCREDI) platform (21 January 2016), emphasizes another aspect that gives reason to devote attention to affordability. The EU has over 4.5 million km of paved roads and 212,500 km of railway lines and has invested about 859 billion euro in its transport infrastructure in the period 2000–2006. The current impact of infrastructure is significant: congestion costs Europe about 1% of GDP every year, and transport is responsible for about a quarter of the EU's greenhouse gas (GHG) emissions. That is one very important subject. However, from a risk elimination perspective, the data are also quite alarming. Failure costs estimated for Sweden and Norway are roughly 10% of total investment costs. For the entire EU, failure costs are about 10–15 billion euro per year. The study of Sweden and Norway also showed that about 30%–50% of failure costs are related to geotechnical matters. Therefore, such costs for the entire EU are more than 4 billion euro per year. Using this data for the Czech Republic, failure costs amount to about 4 billion CZK per year. It follows that there is considerable room for improvement.

Certainly, it is reasonable to ask why geotechnical-related failure costs are so high. Again, the legitimate reply is that geotechnical engineering works with natural materials such as soil or rock, while other structures involve man-made materials such as steel or concrete. This explanation is given already by (Terzaghi, 1959): "Unfortunately, soils are made by nature and not by man, and the properties of nature are always complex."

There are different ways to lower the probability of failure. Some of them are addressed in this publication. However, it is necessary to stress that the principle of structure design takes into account a probability of failure. In fact, this is a principle of the limit state approach to the design of engineering structures.

Probability of the failure of earth- and rockfill dams can be used as an informative example. ICOLD (International Congress on Large Dams) in the 1970s collected information from different countries and for this type of large dam constructed from the beginning of 20th century and evaluated the probability of failure as 1:100 – so it means that of 100 constructed dams, one in reality failed. This corresponds to the probability of 1×10^{-2}. However, with time the probability of failure decreases, mostly as the result of better input data (e.g. hydrological) or of new knowledge accumulating in the branch of geotechnical engineering. For the USA, this probability of failure is now 1:1000 (1×10^{-3}) and discussion is taking place about the possibility of another decrease – to 1:10,000 (1×10^{-4}). Economic evaluation

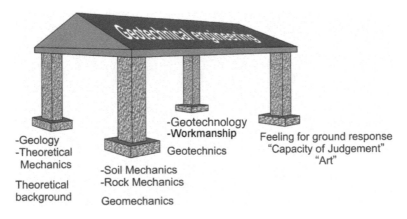

Figure 1.1 Four main columns of geotechnical engineering

is playing a very important role here, as this at the same time means that (theoretically) 9999 dams will be safe, but most of them are designed to very conservative margins. Higher expenses on that side thus markedly exceed the losses caused by one dam failure.

A similar experience was recorded in Czechia where the recommended partial factors of safety applied to the design of spread foundations by the limit state method counted with a probability of failure 1:10,000 (1×10^{-4}) (Vaníček and Vaníček, 2013b). However, after more than 20 years of application, it was concluded that the reality was much better, and that the probability of failure was much lower – approaching 1×10^{-6}.

In order to guarantee not only safe but also economically efficient geotechnical structures, the main approach is by way of knowledge improvement.

One of the plausible perceptions concerning a safe and generally optimal geotechnical structure leads to the conclusion that its success is supported by four columns (Vaníček and Vaníček, 2008; see Figure 1.1). Therefore, the knowledge improvement for the individual columns is a basic assumption of all the processes of geotechnical structure design and construction.

The first column relies on an understanding of natural sciences such as geology, engineering geology and hydrogeology on the one hand, and an understanding of mechanics and the theory of elasticity on the other. This first column can be called theoretical background.

The second column relies on the application of existing theoretical findings to the behaviour of soils and rocks under different stress-strain states and concerns support from soil and rock mechanics more generally on matters of geomechanics.

The third column relies on a combination of theoretical findings with practical technologies and workmanship during the actual execution of geotechnical structures. Therefore this column can be called geotechnics, or geotechnology.

Finally, the fourth column relies on a certain feeling for ground responses to the proposed geotechnical structure, which Terzaghi (1959) declares as "capacity for judgment." He states, "This capacity can be gained only by years of contact with field conditions."

All the aforementioned priorities and aims will be discussed regarding the background to earth structures in transport engineering.

Chapter 2

Risk in geotechnical engineering

The following principle is generally accepted for civil engineering structures. The amount of care devoted to the collection of information, to design and finally to the actual construction is strongly connected with risk. For structures with low risk, this care is much lower than for structures with high risk.

With respect to the assertion made previously, namely that the risk of failure is higher for geotechnical structures than for structures using man-made construction material, we can estimate the differences between individual designers for a simple structural element, as is shown in Figure 2.1 (Vaníček, 2013a). For man-made structures, this simple structural element could be associated with a beam, or for earth structures it might be an embankment. For a steel structural element, the designer defines the properties of the beam that correspond to the calculated loading. The supplier of this beam guarantees the demanded properties and therefore the differences between various designers could be very low, in the order of 3%–5%. For a concrete beam, the supplier guarantees the properties of concrete,

Steel Structures	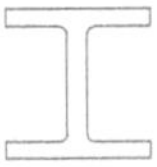	3%–5%	E, σ_d, σ_t
Concrete Structures	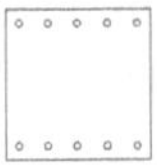	5%–10%	E, σ_d, σ_t
Timber Structures		10%–20%	E, σ_d, σ_t
Earth Structures	 Embankment (borrow pit) 1/1,000,000 from the whole volume indirect methods	cca 50%	E_{def}, ϕ, c, k (based on ρ_d, w)

Figure 2.1 Different levels of uncertainty typical for different structures

and these properties can be checked by additional tests on the supplied concrete or even checked (e.g. with the help of some non-destructive methods) after the application of the demanded structural element. The differences can extend to between 5% and 10%. A similar approach relates to timber structures, but there the properties are influenced also by some irregularities, mostly associated with knag; therefore, the differences can be higher, let us say between 10% and 20%. However, for geotechnical structures, specifically for earth structures, the differences can even be higher than 50%, especially comparing very conservative and very optimistic design approaches, which are strongly affected by our knowledge about the ground and about the fill material. We have to take into account that in such a case we have the chance to test, let us say, only about a one-millionth part (often much less) of the structure volume. Another difference is the fact that the testing of the material properties from the borrow pit, and later on for the constructed embankment, is mostly based on index properties. The quality control of the fill compacted in the embankment is conducted mostly through control of moisture content and dry density (checked by the Proctor test of compaction – w or γ_d), while the design is connected with mechanical-physical properties as shear strength parameters (φ, c), deformation parameters (E_{def}) or a filtration characteristic (k). Therefore, the geotechnical parameters used by different designers can significantly differ. This finally leads us to the conclusion that there is again great scope for improvement.

Risk is directly connected with the probability of failure. Nevertheless, the reasons for any failure can result from the following factors:

a.   The limit state design approach, as based on the theory of probability, counts with a certain (however very low) risk of failure, and this is its basic principle.
b.   Risk of failure is influenced by our level of understanding, and depends on our ability to describe and to understand a very complicated geological environment and to follow-up with a determination of geotechnical properties.
c.   Risk of failure is however also associated with mistakes, in consequence of a lack of education or a lack of sufficient workmanship and control checks.

Point (a) is associated with the fact that even if we follow all recommendations of existing codes, e.g. EC 7, there is a certain risk of failure. Therefore, there is a rightful question addressing what probability of failure is accepted, if and how it can be defined and, if it can, by whom. To this end, the EC 7 Geotechnical Design is giving the national standard body the opportunity to specify partial factors of safety in a so-called National Annex. Some countries did it very carefully, while some others accepted partial factors as recommended in the proposed version of EC 7.

Point (c) is usually under check and control with the help of some regulations. Across countries, these regulations can be different, but in most cases, they are controlled by the Chamber of Civil Engineers, specifying the conditions for persons carrying out geotechnical design and construction work, e.g. a chartered status for geotechnical engineering, which is usually needed for structures associated with higher risk. An expectation that this point can be solved only by the market cannot be fully accepted, especially for problems falling under the subject of environmental geotechnics, as some mistakes can be recognized only after some great or significant time delay.

Point (b) is associated with our level of knowledge. Each generation of geotechnical engineers is trying to understand to the geological environment as much as possible on its level of knowledge. This is similarly valid for determining geotechnical parameters. This means that a model of final calculation representing soil structure interaction is continuously improving with time. The same is true for some aspects of construction technology, which have also significantly improved over recent decades. The risk management process, which considers all uncertainties linked with the ground, can thus recommend the best construction technology. A proposed technology can significantly reduce the probability of failure due to factors that were not discovered or perhaps were neglected during phases of the project preparation.

The presented work deals mostly with this last point and concentrates on expanding and deepening the full scope of our knowledge.

2.1   Risk evaluation

The term "risk" is very general. Some specifications can be found for example on the web pages of Wikipedia. The European project INTACT (impact of extreme weather on critical infrastructures) gives a more focused and refined explication (www.intact-wiki.eu).

Eurocode 7 Geotechnical Design (EN 1997–1: 2004, EN 1997–2: 2006) is valid for European countries, and in part 1 General rules, in chapter 2.1 Design parameters, paragraph 2.1.(8), it states:

> In order to establish minimum requirements for the extent and content of geotechnical investigations, calculations and construction control checks, the complexity of each geotechnical design shall be identified together with the associated risks.

First, it is recommended to separate out geotechnical structures that are associated with negligible risk and for which the minimum requirements will be satisfied by experience and qualitative geotechnical investigation. Briefly stated, this is where all the know-how of our predecessors can be used and relied upon.

For all other geotechnical structures, the minimum requirements should be satisfied by a calculation process, which in the case of EC 7 is based on the limit state principle. However, for geotechnical structures associated with an abnormal risk, alternative provisions and rules should supplement the recommendations and standards specification included in EC 7.

Therefore, to establish optimal geotechnical design requirements, the EC 7 recommends the division of geotechnical structures into three geotechnical categories, 1, 2 and 3. From the previous statement, it is obvious that Eurocode 7 is focused on the conventional types of geotechnical structures, for which more detailed specifications are given.

Boundaries for individual geotechnical categories are not strictly defined, although great scope is given to the individual states accepting EC 7. Therefore, the authors of the final version of EC 7 are putting forward some more detailed specifications in additional publications (Frank *et al.*, 2007, 2011; Bond and Harris, 2008). For example, Frank *et al.* recommend incorporating such geotechnical structures into the third geotechnical category, for which at least one criterion falls into the highest scale of evaluation, in other words:

- Large and atypical structures;
- Structures associated with abnormal risk;
- Structures which are interact with atypical or rare foundation conditions;
- Structures which are loaded atypically or abnormally.

On the other hand, the EC 7 indicates the possibilities on how to determine the risk with which an individual geotechnical structure is associated, as in:

A.   *Direct specification of geotechnical structures falling into different geotechnical categories.*

Here, underground structures (tunnels) can be used as an example. EC 7 specifies that tunnels in hard rock fall into the second geotechnical category (GC). Therefore, it means that all others are falling into the third GC. Nearly all large dams fall into the highest GC, as instead of the general rules expressed in EC 7, some additional conditions specified e.g. by ICOLD (International Congress on Large Dams), should be followed.

B.   *Risk evaluation based on the specification of:*
  - Complication of ground conditions (complication of geological and geotechnical conditions);
  - Demanding nature and level of geotechnical structure;
  - Impact of failure of geotechnical structure on the environment (the so-called consequence classes).

C.   *Risk evaluation on the specification of uncertainties with which the main phases of the design and construction of geotechnical structures are associated.*

More detailed specification of the possibilities for B and C are presented, insofar as the process of risk evaluation for both cases makes it possible.

2.1.1   Risk specification taking into account ground conditions, type of structure and impact of failure on the environment

The simplest way to specify any risk is to divide individual factors into three levels. Ground conditions (geological and geotechnical conditions) can be:

- Simple, modest or very complicated.

Geotechnical structures with respect to the level and nature of their demands can be:

- Unpretentious (simple), moderate (conventional) or very demanding.

The impact of structure failure on the environment generally is indicated by EN 1990 (EC 0), which uses the term "consequence classes" with respect to the impact on human life, society and the environment. The impact of any failure based on the first generation of EC 0 can be divided into:

- Practically negligible, insignificant;
- Moderate;
- Very high.

By a mutual combination of these three factors, a differentiated evaluation is possible (see Table 2.1).

Table 2.1 Proposed levels for evaluation of structure, ground and the impact of failure to give a total risk evaluation (for classification into GC)

Structures	Simple (unpretentious)	Middle (conventional)	Very demanding
Ground	Very simple	Modest	Very complicated
Impact of failure	Practically negligible	Moderate	Very high

However, there is a general agreement that geotechnical structures falling into the first GC have, for all the aforementioned factors, the lowest level of classification and the geotechnical structures falling into the third GC have at least one factor falling into the highest level of classification.

The geotechnical structures falling into the second geotechnical category have therefore different types of combinations, which is why the EC 7 is focused in this direction. To be able to distinguish between different structures falling into the second GC, some proposals were made to further subdivide the category. For example, one proposal suggested three subdivisions for the design of a specific structure, according to the different demands on the quality of soil samples and the tests performed on them (Vaníček, 2016).

The recommended evaluation into five different categories is in principle comparable with the proposal for a second generation of Eurocodes, where five consequence classes (CC) are recommended – CC0 to CC4. However, the Eurocodes do not cover cases with zero impact (CC0) or alternatively with extremely high impact (CC4). According to prEN 1997–1:20xx (E) Draft April 2018, consequence classes CC1–CC3 are combined with a different geotechnical complexity class (GCC), where this GCC involves the classification of a geotechnical structure based on the complexity of the ground and ground structure interaction and where three GCCs are distinguished. The result of this combination is the division of geotechnical structures into three geotechnical categories, so this means that the original classification into three GCs was retained.

The initial qualification and identification of geotechnical structures for these individual geotechnical categories should start during the earliest phases of project preparation and should become clearer over time.

2.1.2 Risk specification taking into account four main phases of a geotechnical structure design and execution

In this case, the risk specification is based on detailed evaluation of the uncertainties with which the main phases of the structure design and execution are connected. This specification is the more general. However, it still allows deeper insight into all aspects of the problems associated with geotechnical structure design and execution. Therefore, the individual phases will be discussed in more detail both in this chapter (from the viewpoint of risk specification) and in Chapter 3 (but from the viewpoint of risk reduction).

Here it is very useful to state that basic principles, the design of the engineering structure, and their execution are all linked. The designer has the main responsibility for a safe and optimal design. The designer should take into account all uncertainties, which will be briefly discussed in this section, and subsequently call on them when working on the geotechnical design report. But the designer is not the only person responsible for any risk of a structure failure or loss of its serviceability. Other partners share this risk, namely the person

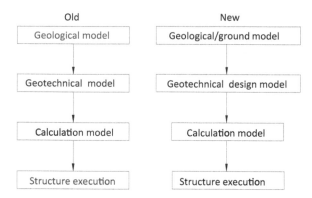

Figure 2.2 Main phases of geotechnical structure design and execution

responsible for the ground (geotechnical) investigation report (GIR), the person responsible for structure execution (the contractor) and finally, but quite importantly, the investor (the owner). Only a very high level of cooperation among all these people can bring the expected results.

Since different people are responsible for them, the following four main phases of the whole process are mentioned (Figure 2.2) when these phases are also underlined in both first and second generations of EC 7.

2.1.2.1   Ground model

The Ground/Geotechnical model (GM) is part of the Ground (Geotechnical) investigation report (GIR) and consists of two main parts, a geological model and an overview of tests performed on the ground, either in the field or in laboratory tests.

A geological model provides in 2D or 3D a substitute visual representation of the real geological environment. A geological model specification is time-dependent process, as over time it improves from the conceptual level up to the model used by the designer. Later on, it can be clarified in more detail when real conditions are detected during the phase of structure execution.

The uncertainties linked with the geological model (or its credibility) strongly depend on:

• Complexity of the geological environment;
• Actual state of exploration of this geological environment;
• Extent of the ground investigation and its quality;
• Ability and professional skill of the persons responsible for the site investigation and interpretation.

The last point is closely related to column 4 in Figure 1.1. The geological model is not only a geometrical interpretation of the real environment. This model should also be interpreted and supplemented by the expected interaction of this geological environment with the proposed structure.

Experienced engineering geologists can specify this interaction either from the limit state point of view (how the structure is influenced by ground conditions) or from the environmental view (whether the proposed structure can have a negative impact on this geological

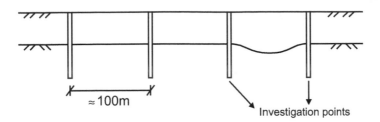

Figure 2.3 Uncertainties in a geological model caused by a great distance between investigation points

environment), as the final evaluation can be used also for the environment impact assessment (EIA) process.

An additional note focuses on the extent of a ground investigation. Close cooperation is needed between engineering geologist, designer and owner. Most geotechnical engineers will agree with the following statement. The total construction cost can be significantly reduced by spending slightly more of the budget on ground investigation and laboratory and field tests in order to capture the subsoil conditions more precisely. However, the question concerning an economically optimal budget is duly justified and has to be kept in mind.

The aforementioned problem can be shown for 2D structures such as earth structures of motorways and railways. A very important question relates to the distance of the investigation points. For a larger distance, the interpretation of the geological model (the geological profile in longitudinal direction) is more complicated and some irregularities can be missed, e.g. Figure 2.3. Being able to specify the optimal distance is therefore a very sensitive problem to which all partners should propose solutions.

The last note focuses on the difference between the embankment and cut from the view of the ground profile control during structure execution. The ground profile for the cut can be (and should be) controlled, either for the excavated slopes or for the bottom of the excavation, while the control of the ground profile under the embankment is not performed.

FIELD AND LABORATORY TESTING

The main aim of this section is to specify geotechnical data for each lithological layer of the ground or for discontinuity between these layers, also taking into account fluctuation of groundwater.

The uncertainties associated with this section depend on similar factors as for the geological model. However, a widely experienced specialist on laboratory and field tests, with a good background in soil and rock mechanics, plays an important role here. The credibility of the results also depends on an appropriate selection of the individual tests.

What is predictive ability of the test results for the control of limit states? – particularly for calculation model, which can be as analytical or numerical one with demands for different input data. Therefore, the designer of factual geotechnical structure should recommend or at least oversee proper selection of test and input data. Some geotechnical data have great value for the contractor, e.g. for an optimal selection of construction technology, which can have a close relationship to the bidding process. Classification of ground for earthworks is also very important for earth structures in transport engineering.

The results of field and laboratory testing are generally presented in the Ground (Geotechnical) investigation report (GIR) in tabular form, with a summary of all results for individual lithological layers or discontinuity. This form of presentation should be supplemented by an interpretation of results, namely with respect to:

- Comparison of the obtained geotechnical data with existing data obtained for a similar type of ground in the past. Whether they fall inside the range of expected (standard table) values or are outside this range is seen.
- Comparison of obtained data regarding groundwater with the expected fluctuation.
- The form of specification of geotechnical parameters. These parameters can be acquired from the test results directly, mostly from lab tests. Such parameters are denoted as measured values. But the geotechnical parameters can also be acquired from indirect tests, e.g. from the field tests as penetration and pressiometer tests. Such parameters after that are denoted as derived values. The derivation method used should be mentioned, as should the theory, correlation or empiricism used.

2.1.2.2   Geotechnical design model

The Geotechnical design model (GDM) is closely connected with the ground model. Both are continuously evolved as the design proceeds, and the individual steps of GDM development can be used for the individual steps of the design. The final version of the GDM specifies the characteristic values of the geotechnical parameters for different lithological layers of the ground or for discontinuities between them. These characteristic values should be selected as a cautious estimate of the value affecting the occurrence of the limit state and are used subsequently for the calculation.

The comparison of both aforementioned models (GM and GDM) will show very clearly what values were selected for the calculation from values obtained in the phase of investigation. This comparison is very important for the control phase. On the one hand, it shows how conservative or optimistic the designer was, and on the other hand, how close the selected values are to the values obtained on samples from exposed ground during structure execution (ground excavation).

The specification of the characteristic values is the most sensitive part of the whole project. By this phase, the geotechnical design differs from the design of other civil engineering structures. Two designers can differ significantly.

The existing version of Eurocode 7 from 2004 refers to two ways by which characteristic values can be selected:

- Standard tables of characteristic values related to soil investigation parameters (standard tables of strength and deformation properties based on soil index properties and on soil classification).
- Statistical methods with statistical evaluation of the results obtained for an individual layer from lab or field tests.

In both cases, the characteristic value shall be selected as a cautious estimate of the value affecting the occurrence of the limit state, complemented by well-established experience.

The exploration of these two ways depends on:

- The phase of ground investigation;
- The risk with which the structure is connected.

It is obvious that standard tables can be utilized for structures connected with low risk or at the end of the first step of ground investigation (after desk study and site reconnaissance), when preliminary soil classification is possible. Statistical methods, on the other hand, are typical for structures connected with high risk or can be applied at the end of the design investigation phase, when for each layer there is a sufficient number of measured results. From the above parameter selection alternatives, it is obvious that a combination of both ways is typical for geotechnical structures associated with medium, moderate risk or after the preliminary phase of ground investigation, when the number of results of strength and deformation properties is insufficient for statistical evaluation.

2.1.2.3 Calculation model

The calculation model is the most frequently used method of limit state verification. The calculation model will describe the assumed behaviour of the ground for the limit state under consideration. Nowadays, two basic calculation models are used:

- Analytical model.
- Numerical model.

The analytical calculation model divides control into two basic parts – control of Ultimate limit state (ULS), which is the limit state of failure, and control of the Serviceability limit state (SLS), which is mostly associated with deformation.

At this point, it is suitable to present at least a short note on the principle of the limit state design. The characteristic values of the geotechnical parameters represent a certain value selected with respect to the solved limit state. The characteristic value of deformation properties for the SLS will be close to the average value, as the result of the deformation calculation should be as close as possible to the consequently measured value. On the other hand, the characteristic value of the strength properties will be somewhat conservative and will be influenced primarily by:

- Length of the slip plane along which a given geotechnical structure can fail. For a longer slip plane, there is a higher chance for a compensation of variation of shear strength on both sides. The selection of a characteristic value for the shorter slip plane will be more conservative.
- The ability of the geotechnical structure to transfer loads from weak to strong zones in the ground.

However, for the calculation of limit state of the ULS type, the characteristic value (X_k) is mostly not used directly, but the design value of geotechnical parameter (X_d) is used, when the relation is:

$$X_d = X_k/\gamma_M \qquad (2.1)$$

Where γ_M is the partial safety factor for a material property.

The design is fulfilling the ULS limit state when the equilibrium on the slip plane is reached.

The numerical calculation model examines the stress-strain state of the ground, which is in some interaction with the proposed geotechnical structure. For the observed element the change of stresses or deformation can be obtained for a given change of loading. Nevertheless, even then it is appropriate to solve two basic limit states (ULS and SLS) separately, when the characteristic values are applied for solving the SLS while the design values are preferred for the ULS.

The uncertainties of the analytical model are relatively high, and for each typical geotechnical structure or some part of it, there should be a separate definition. For example, for the problem of slope stability, the calculation model should take into account not only the geological model, but also:

- Seepage and pore water distribution;
- Short- and long-term stability;
- Type of failure (circular or non-circular surface, toppling, flow).

The appropriate selection of the slope stability method is very important, as each has different basic assumptions (e.g. the Bishop or Janbu methods etc.). Therefore, the selected method should be well set out together with basic assumptions.

The uncertainties of the numerical method (mostly Finite Element Method (FEM)), or of the risk, are connected with the credibility of the substitution of a real zone by the geological and geotechnical design models with finite elements. Generally, this risk is associated with:

- Correct and precise division of the observed zone (geotechnical design model) into individual elements;
- The function expressing the change of properties within individual elements – basic function;
- The constitutive model, which is expressing the dependence of the deformation changes on stress changes, and finite elements on model structural components;
- Finite elements to model interfaces;
- Boundary conditions;
- Ability to model technological sequences.

Therefore, the selected numerical model should always be closely specified, at least for the possibility to control the input data of geotechnical parameters and for the aforementioned points respectively, for the possibility of controlling the presented results. In general terms, the model should be validated.

More specifications concerning calculation models for earth structures can be found in Vaníček and Vaníček (2008) and will be raised also in Chapter 3, mainly concerning how the uncertainties (risk) can be reduced.

2.1.2.4 *Geotechnical structure execution*

Risk connected with geotechnical structure construction has two different levels:

- The technical level, which will be discussed in Chapter 3.
- The legislative, juridical level.

This second level is mostly associated with implementation of new geotechnical structures in the vicinity of, possibly also in close interaction with, older existing ones.

All we really know is that each change in stresses initiates changes in deformations. Therefore, if stress changes in the ground induced by a new structure also affect the ground below the older structure, these stress changes must also create deformations below this older structure. However, the owner of the older structure usually agrees with the new one only under the condition that "the new structure will not have any influence on the older one." This condition is an obvious contradiction. However, it is accepted and transformed by the designer and the contractor into a new condition, namely that "the change does not cause visible deformations – cracks." Consequently, both the design and construction technology adopt this new condition. Special care should be devoted to neighbouring historical structures that are usually much more sensitive to the additional deformations. As a certain protection, the passportization of the old visible cracks on an existing older structure can be done beforehand.

However, there is another sensitive problem. The owner of the older structure might allow some provisional actions to guarantee the safety of both structures during the construction period. However, after that, there can be a demand to deactivate these provisional actions. For example, this could be for anchors situated below the older structure. A typical case in this direction is the collapse of the twin towers of the World Trade Center in New York. The excavation of the ruins was significantly held back while the stability of external walls was restored.

A short additional remark should be made dealing with the lowering of groundwater. This lowering can have a negative impact on neighbouring structures as it is at the same time increasing effective stresses. The increase of effective stresses below the older structure leads to additional settlement, mostly likely a differential one. The calculation has to evaluate this risk and propose some counter measures, if needed.

2.1.3 Total risk evaluation for earth structures in a geotechnical design report

The geotechnical design report summarizes all the collected information, presents an evaluation of risk (all the uncertainties) for the main phases of the design process and selects an approach for reducing this risk. Figure 2.4 shows the sequence of these four main phases for a typical case of earth structures in transport engineering, in this case embankment construction (Vaníček, 2016). They are part of the Geotechnical design report (GDR), of which a Geotechnical (Ground) investigation report (GIR) is the important first component.

From this schematic figure, it is obvious that the geological model has two main parts: a geological model of the ground and a geological model of the borrow pit from which the construction material is used. In the first instance, we are obtaining basic information about the ground with which the proposed structure will be in interaction, and in the second instance basic information about the soil that will create this proposed structure. For the first, the tests are directed on the mechanical-physical properties needed for the control of limit states. For the second, however, the tests are directed primarily on technological aspects, involving compaction tests, e.g. the Proctor standard test. The optimal moisture content is determined and compared with the result of natural moisture content in the borrow pit – to see if any adaptation in this direction is needed. Subsequently, mechanical-physical tests are performed on samples compacted in the laboratory to get the recommended range of acceptable moisture content.

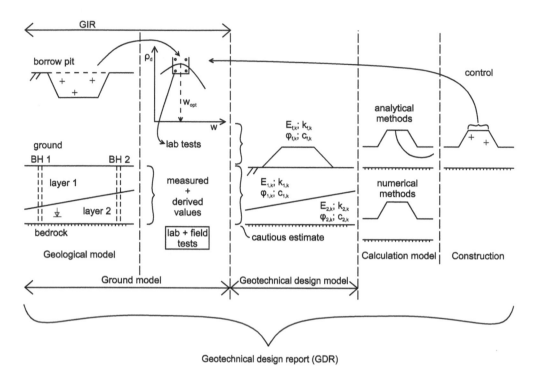

Figure 2.4 Sequence of the main phases influencing overall risk within the logical scheme of the GDR for earth structures

The Geotechnical (Ground) investigation report should contain, to put it in brief and practical terms, sufficient data concerning the ground, borrow pit and groundwater conditions at and around the construction site to give an accurate description of the essential ground and borrow pit properties:

- Geological models for ground and borrow pit;
- Tabular and graphical presentation of the tests results carried out both for subsoil and embankment;
- Detailed description of the construction site from the engineering geological and hydrogeological point of view, stressing all relevant problems which can be associated with this locality and the proposed earth structure.

Subsequently, the designer of the geotechnical structure defines, with a cautious estimate, characteristic values of the geotechnical parameters, where the main accent is placed on values affecting the occurrence of the limit state. The geotechnical design model consequently indicates these selected characteristic values of the geotechnical parameters. A special case occurs when the designer suggests a ground and/or embankment improvement. In that case, the selection of the characteristic values of the geotechnical parameters is more complicated, as the designer should use some values for the recommended improved ground in advance, with limited possibility to control them.

In the next step, the designer selects the most relevant calculation model, most often selecting between analytical and numerical models. Using this calculation model, the designer should confirm that all relevant limit states are fulfilled.

Consequently, the designer defines the conditions for the technology, where the compaction specification plays the most important part. In the case of a proposed ground improvement, the specification is also in this direction.

Finally, the designer should also include for the GDR a plan of supervision, monitoring and structure maintenance, as is appropriate. Earth structure quality control, and the control of soil compaction, is again a priority.

The contractor, during the phase of construction, should attest that all the demanded requirements were fulfilled. With respect to the compaction control by the standard procedure, the moisture content, together with dry density, are monitored and checked to affirm that they are falling into the required range.

Earth structures are, from this view, untypical as to classical standard checking procedures. The embankment quality control is indirect, via index properties, while mechanical-physical properties, which were applied during the limit state design, are not controlled. Paradoxically, the real properties of the subsoil during the phase of earth structure construction are not controlled at all, with a small exception for cuts. Therefore, the possibility of control, if the values used during the phase of design are satisfied, is very limited, especially when compared with other structures, where man-made construction material is used. Consequently, the earth structures of transport engineering are different from other geotechnical structures, while for foundation structures and underground structures, the real conditions are at least partly disclosed.

As was mentioned before, the comparison of measured values of the geotechnical parameters during ground investigation with characteristic values used for the design, or with values discovered during the construction phase (or again during a supplementary geotechnical investigation), is very useful with a high predictive significance or meaning. This is not only important for the solved case but also for subsequent similar cases.

2.1.4 Risk and natural hazards

One of the different possibilities for how to quantitatively express potential risk is the equation expressed below, e.g. Kalsnes *et al.* (2010). This equation is most often used for risk evaluation associated with natural hazards.

$$R = H \times V \times E \tag{2.2}$$

Where:
- R is risk associated with a particular danger (threat) – such as floods, landslides, rock falls, avalanches, earthquakes, volcanic activity, hurricanes, tornados, tsunami etc.;
- H is hazard – the probability that a particular threat will occur within a given period of time;
- V is vulnerability – the degree of loss to a given element within the area affected by this threat, when V = 1 for total loss and V = 0 for zero loss;
- E is the expected cost of a total loss of elements at risk.

Over at least the last century, there are relatively well-documented cases which allow one to find the relationship between the frequency of a certain threat and how it corresponds to a certain total loss (this can be expressed either in monetary cost or fatalities), e.g. Proske (2004) and Briaud (2013). A typical and well-known relationship concerns the evaluation of floods. Floods with a statistical frequency of only once over a hundred-year period are likely to cause a much higher level of damage than floods with a statistical frequency of five years. However, these relationships are typically expressed for all potentially affected areas, not only for transport infrastructure. Chapter 5 will discuss some problems associated with natural hazards, especially measures that can help to reduce loss of life and the material damage associated with them. With respect to risk management, preventive measures are directed at:

- Better forecasting of an individual event;
- Strengthening the resistance of structures;
- Producing a methodology on how to behave during such an event, particularly when it happens during structure construction.

The last point is very sensitive when an area through which the proposed new transport infrastructure is identified during the phase of investigation as landslide or flood prone. Cooperation of all partners is a very delicate issue. For example, for a flood-prone area, the contractor has to protect the construction site for a certain type of flood (e.g. for a 10-year or a 25-year flood), and the question arises as to how this problem is covered in the bidding conditions specified by the investor.

Other natural events, such as heavy snowfall, also affect transport activity (using the criterion of availability), but for a limited period only without any special impact on the earth structures of a transport infrastructure.

In some areas, the first event can trigger another. The extreme case is an earthquake which can trigger landslides or rock falls, which can then block a river valley. An increasing water level behind a naturally constructed "dam" can cause its failure with heavy floods below it.

Chapter 3

Geotechnical risk reduction during earth structures of transport engineering design

Safe and optimal design of earth structures of transport engineering (TE) is a relatively challenging process. Among basic points of this process belongs:

- Complying with all potential risks – uncertainties – that were discussed in Chapter 2. The primary aspiration is to decrease the range of uncertainties with which the design for an individual structure is connected.
- Complying with the sequence of individual steps, mainly with respect to the successive refinement of information about the ground.
- Adhering to different rules and standards during the site investigation phase, e.g. with respect to quality of samples and test performance, while simultaneously using a personal approach to their evaluation, namely during the selection of characteristic values of the geotechnical parameters.
- Selecting the most appropriate methods for assessing structure safety, when the main attention is devoted to the limit state approach.

3.1 Basic principles of the geotechnical structure design

The classical approach to the geotechnical structure design based on calculation model was to apply a total factor of safety (stability), denoted most often as F or FoS. Detailed specification of FoS was different for different geotechnical structures. For a slope stability problem, the factor of safety was expressed as a ratio of forces resisting sliding to forces eliciting (disturbing) sliding on an expected slip surface. The ratio of bearing capacity of foundation to contact pressure under the footing bottom was applied for spread foundations. For a gravity retaining wall, the factor of safety against wall overturning is expressed as ratio of two moments to the point of overturning. Demanded total factor of safety is most often in the range of 1.5 (for slope stability) to 2.5 (for spread foundations). The total factor of safety covers all uncertainties, mainly soil parameters and loading.

Subsequently, the deformation was calculated, when the settlement of the spread footing is very important. Deformation of the transport infrastructure surface (road pavement, rail vertical alignment, runway surface) is also very important, mainly with respect to differential settlement. Demands on this differential settlement are also variable, with a great difference for local track and for the track for high-speed trains. Initial differential settlement also affects surface maintenance or even service life.

Within the last 50 years, there has been a continuous diversion from the total factor of safety to design based on the principle of limit states. In the Czech Republic, this process started in 1967 (Vaníček and Vaníček, 2013b). The first proposal of the European code geo-technical design, prepared by a team under the leadership of Prof Ovensen, elaborated this approach for all geotechnical structures in 1986. All uncertainties are differentiated mainly on material and loading (action) and covered by partial factors of safety γ_M on material properties, γ_F on actions and γ_R for resistance (material strength). Different combinations of these partial factors enable one to define different design cases.

This approach differentiates between two basic limit states:

1.   Ultimate limit state (ULS);
2.   Serviceability limit state (SLS).

When the following inequality for the verification of the first – verification of the ultimate limit state is used:

$$E_d \leq R_d \tag{3.1}$$

Where:
E_d is the design value of the effect of actions;
R_d is the design value of the corresponding resistance.

Briefly speaking for the slope stability problem: When applying a partial factor for material γ_M on characteristic values of shear parameters, the resistance will still be higher than the design value of the effect of actions (disturbing actions on slip surface). The second generation of the Eurocode 7 Geotechnical Design (EC 7) denotes this approach as the material factor approach (MFA).

For the verification of the serviceability limit state, another inequality is applied:

$$E_d \leq C_d \tag{3.2}$$

Where:
E_d is the design value of the effect of actions (calculated deformation);
C_d is the limiting design value of the effect of an action (max. acceptable deformation).

More specifications will be shown later in this chapter for specific problems.

The previously discussed approach to the verification of limit states based on calculation – or calculation model – is the principle applied most often. However, there are also other possibilities.

The *design with the help of prescriptive measures* is used for geotechnical structures with very low risk – those with practically zero consequence of failure. This approach is therefore typical for small structures. The case of low embankment (e.g. up to 3 m) situated on well-known ground is such an example. However, this approach can also be applied for cases where there are no calculation models, as is the case, for example, of the design of a filter for internal erosion protection.

The *design with the help of geotechnical structure behaviour modelling* is typical for new types of structures, for new geotechnologies, with little practical implementation experience. Most appropriate is the model scale 1:1, modelling real structure. (Boden *et al.*, 1979) describe such a full-scale trial model of reinforced earth walls. By this model, TRRL (Transport and Road Research Laboratory) in the UK started to investigate the methods of design

Figure 3.1 Model scale 1:1. Testing of actual response of railway embankment during high-speed train passage

Source: Courtesy of Pardo de Santayana.

and construction of reinforced earth walls. Testing of the actual response of railway embankments during high-speed train passage started in Spain – in the *Laboratorio de Geotecnia* of CEDEX Madrid (Estaire *et al.*, 2017), see Figure 3.1.

Physical models with smaller scale have a number of distinct advantages over full-scale tests. However, there is a very sensitive problem connected to the model scale. Model tests on centrifuge are much less sensitive to the model scale (Schofield, 1980; Taylor, 1995) and therefore are now preferred.

Probably the best idea is to perform an *in situ* test on the locality where the geotechnical structure is proposed. De Beer (1969) describes such a case. The fundamental question was to design the slopes of excavation in such a way that they should be stable over a period of about one year. The total depth of excavation was nearly 30 m, with the lower part in Boom clay. As the designed cut had a length of around 1 km, test pits bordered by slopes with different inclinations were realized. The results of the inclinometers, piezometers and horizontal deformation measurement of the slope top helped to design optimal inclination, which guaranteed a stable state for the demanded period.

Another way to design the geotechnical structure is to apply the *observational method*, which is described in more detail in section 3.4.6. The observational method is suitable mainly for deep cuts of transport engineering. The optimal soil inclination depends on slope monitoring, similar to the previously described case. Another application is for a high embankment situated on soft ground, particularly when inclined. The speed of measured pore pressure increase and subsequent dissipation in the soft ground influences the speed of embankment filling.

However, the basic principle – for all basic types of geotechnical structure design – is the same. The structure should be designed and constructed in such a way that during its assumed service life, it should:

- Resist – with demanded degree of safety and economic efficiency –all load reactions and all impacts that could, with the required probability, occur during performance and utilization.
- Serve the given purpose.

3.2   Geotechnical/Ground model

Recent experiences have shown that the total construction cost can be significantly lowered by spending slightly more on field investigations and laboratory soil tests in order to determine the subsoil conditions more precisely. Those efforts make it possible to propose a more rational design as well as better construction planning (Towhata, 2017). Therefore, the risk sharing between individual partners of the construction process can lead to a decrease of risk (Vaníček, 2013b). On the other hand, there is no reason to spend more money on field investigation for a simple case, when the test results would have very little influence on the design, or when the test results would have low predictive value for the limit state design.

Basic regulations and recommendations about ground investigation, particularly in Europe, are parts of EC 7 Geotechnical Design, specifically EC 7–2 Geotechnical/Ground investigation (see e.g. Frankovská, 2019). EC 7 ascribes great importance to the information about the ground, as evidenced by the statement, (EN 1997–1: 2004, 2.4.1 (2))

> It should be considered that knowledge of the ground conditions depends on the extent and quality of the geotechnical investigations. Such knowledge and the control of workmanship are usually more significant to fulfilling the fundamental requirements than is precision in the calculation models and partial factors.

3.2.1   Geological model

A geological model either in 2D (plane) or 3D (space) is, in its planar or three-dimensional modifications, always more or less a substitute for the real geological environment. Two examples for simple and complicated cases are presented in Figure 3.2.

Therefore, a specification of geological model is time-dependent process as with investigation stages is gradually complemented. Geotechnical/Ground investigations (GI) shall be planned in such a way as to ensure that relevant geotechnical information and data are available at the various stages of the project. The various stages should be able to answer questions raised during planning, design, and construction of the actual project.

The steps taken for ground (geotechnical) investigation for important structures that fall into the higher geotechnical category (GC) consist of desk study, preliminary GI, design

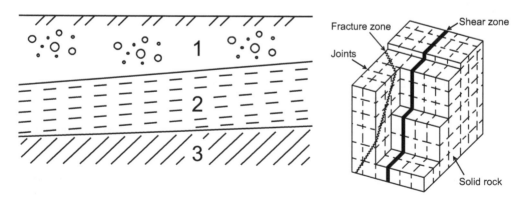

Figure 3.2 Geological models for simple (2D) and complicated (3D) cases

Table 3.1 Individual steps of ground investigation and geotechnical design – geotechnical design report (GDR) and geotechnical construction record (GCR)

Ground investigation (GI)	Geotechnical design
Desk study	Study – investment project (idea)
Preliminary GI	First step project – documentation for planning inquiry
Design (detailed) GI	Design (GDR) – documentation for building permit
Supplementary GI	Design modification – for construction
Confirmatory GI	GCR – documentation for real structure performance

(detailed) GI with GIR, supplementary GI and observational (confirmatory) GI. Table 3.1 illustrates the interrelationship of the individual steps of GI with individual steps of geotechnical design. It is obvious that for geotechnical structures with low consequence classes, respectively low geotechnical category, the number of these steps can be significantly reduced.

Desk study uses all existing information supplemented by site inspection. Different maps compose the basic initial source of information, starting with geological, engineering geological and hydrogeological maps. Similarly, the database of existing investigation points is a very important source of information – with results of previously performed GIs, especially in close vicinity of the proposed geotechnical structure.

Preliminary GI already exploits investigation methods, although only on a limited scale that serves as confirmation of assumptions based on the desk study.

Design (detailed) GI is the most important stage, as its output – the ground (geotechnical) investigation report, is directly used for the design – the geotechnical design report.

Supplementary GI comes into question when during construction some differences in the ground model were registered and design modification is needed, namely from the construction technology point of view. Changes most often also require recalculation with respect to the design situations.

For most of the geotechnical structures, but not for all, there is a chance to confirm an expected situation expressed in the ground (geotechnical) investigation report. This phase is therefore denoted as *confirmatory GI*. For earth structures of transport engineering, this possibility is typical for cuts. Outcrops either on the slope or at the bottom of a cut allow for this comparison, not only for the geological model but also for the test results subsequently realized for individual layers of the outcrops. Regrettably, this chance is typically not valid for construction of an embankment, as subsoil is typically not uncovered.

The results of a confirmatory GI serve as important chance to define the difference between the results of this last stage of GI with the results stated in the Ground investigation report (GIR), which were known during the design phase. At the same time, they can permit their comparison with characteristic values of the geotechnical parameters, to what extent the designer during their selection (cautious evaluation) was being optimistic or conservative. The results are then part of Geotechnical construction record (GCR) serving as the final documentation for the realized structure.

As was mentioned previously, the design (detailed) GI phase is the most important phase, and therefore some generally valid specifications of ground investigation will be shown for it. In this case, the ground investigation should comprise at least:

- Drilling and/or excavations (test pits including shafts) for sampling with subsequent laboratory tests;
- Groundwater measurements;
- Field tests.

Detailed descriptions of subsurface exploration are part of different books on engineering geology, starting with Záruba and Mencl (1976). The main aim of this exploration is to receive a very good picture about the ground profile, with distinct boundaries between individual lithological layers. Subsequently for each layer it is recommended to obtain samples for follow-up laboratory tests. It is also obvious that for this part of the design process, the relation between care devoted to the ground investigation and the level of geotechnical category is important. In this direction, closer specification is part of EN ISO 22475–1 Geotechnical investigation and testing – sampling methods and groundwater measurements. This code was prepared by the European Committee for Standardization (CEN), Technical Committee CEN/TC 341, Geotechnical investigation and testing with following scope: standardization in the field of geotechnical investigation and testing pertaining to equipment and methods used for drilling, sampling and field and laboratory testing.

With respect to *sampling*, EN ISO 22475–1 considers three sampling method categories, depending on the desired sample quality:

- Category A sampling methods. These methods are focused on the best quality samples, in which no or only slight disturbance of the soil structure has occurred during the sampling procedure or in handling of the samples. Mechanical-physical properties are tested on such samples, as they are needed for the design of structures falling into GC with high risk, typically GC 3.
- Category B sampling methods. These methods contain all the constituents of the *in situ* soil in their original proportions and for the fine soil to retain its natural water content for specification of its consistency. Index properties are realized on such samples, enabling the soil classification. Subsequently the characteristic values of the geotechnical parameters can be determined from standard tables based on soil classification. They are therefore appropriate for GC with low risk.
- Category C sampling methods. The soil structure in the sample has been totally changed. The general arrangement of the different soil layers or components has been modified so that the *in situ* layers cannot be identified accurately. The water content of the sample needs not represent the natural water content of the soil layer sampled. Such sampling methods can be used only for simple geotechnical structures, with negligible risk and where ground conditions are well known, or for testing on reconstituted samples.

As the sampling category depends on the geotechnical risk, respectively on geotechnical category, this should be specified in the early stages of the investigation, in general at the end of the preliminary GI. The designer is responsible for final specification of the geotechnical category, although close cooperation of all construction partners is needed, particularly from the side of the person responsible for the geotechnical investigation report. Laboratory tests of mechanical-physical properties should be carried out under conditions which are expected *in situ* – as hydraulic gradient, range of stresses, range of relative deformations, for more important problem under stress paths simulating the expected ones. The output from lab tests are measured results of the geotechnical parameters.

With respect to *groundwater measurements,* two basic opportunities are there: open or closed groundwater measuring systems. *Open system* is suitable for ground with high permeability, and groundwater level is measured in an observation well by piezometer. On the other hand, *closed systems* are appropriate for all types of soil or rock. A pressure transducer at the selected point directly measures groundwater pressure. Therefore, the measuring point should be adequately sealed off from other layers or aquifers.

Knowledge of groundwater pressure is very important for earth structure safety, primarily for the initial state (before the design). However, knowledge about possible fluctuation of groundwater pressure during structure lifetime expectancy is very important as well. The maximum (or minimum) expected level based on historical data should be part of the investigation report.

For structures with higher risk, it is recommended to measure groundwater pressure both during and after the construction. Therefore, it is proposed to leave the installed piezometers throughout the project construction until the phase of confirmatory GI. After that, the designer has a chance to compare expected groundwater pressure with reality. This practice is typical for the observational method of structure design.

Field tests have been increasingly performed in recent decades. Advantages of field tests are in their speed and economy, although sometimes they are limited by geological profiles and ground characterization, particularly for course soils or for ground containing large boulders. Nevertheless, in some countries, where the conditions are very similar within a wide area (e.g. where fine sediments prevail), the application of field tests is very efficient and based on updated experiences with high predictive value. Nevertheless, combining lab tests with field tests is the best way to eliminate the drawbacks connected with each way alone. Core drilling (plus sampling and testing), for example, is supplemented by a dynamic penetration test. Results of penetration tests can confirm the geological profile between individual boreholes. The range of field tests is very wide, and some overview is given by Mayne (2012) and Briaud (2013), for example, or with closer specification on individual type of tests (cone penetration testing) by Lunne *et al.,* (1997). Roughly, the distinction between field tests is as follows:

- Penetration tests – measuring penetration resistance of the ground profile (e.g. dynamic penetration, static penetration, standard penetration test etc.).
- Tests for borehole widening (e.g. pressure meter test, test with flat dilatometer).
- Test measuring shearing resistance – as vane test.
- Plate load tests – static and dynamic.

EN 1997 and EN ISO 22476 indicate these tests, as after their evaluations some geotechnical data can be interpreted. In some cases, the results are used for direct design of individual geotechnical structures.

Final summary of the field tests. A great advantage of the majority of the field tests is to provide a continuous vertical geological profile. First, it is a quick and cheap determination of boundaries between individual lithological layers. Very often these field tests are used as supplementary tests for the control of boundaries obtained with the help of sampling, especially if the construction site covers a large area. Second, we are obtaining continuous changes for individual layers and as such, the number of results is sufficient for statistical evaluation. The last advantage is connected with a very good determination of density for coarse soils, which is a great problem when only sampling is used. In most cases, the geotechnical parameters needed for the design are not measured directly. They are derived from the measured values with the help of theory, correlation or empirical evaluation. Therefore, they are called derived values of the geotechnical parameters. Correlation and empirical evaluation are often valid for a given geological condition and local practice, therefore the first application of field tests in new locality (country) needs a certain precaution.

One of the most sensitive questions with respect to the ground model is the *range of the geological environment, which should be investigated.* The proposed range depends not only on the proposed geotechnical structure, but also on the ground plan of the structure and on an active depth (zone of influence) up to which changes in stresses can still have significant impact.

Risk resulting from far distant boreholes situated along the longitudinal axis of the transport infrastructure was already mentioned. Field tests or geophysical investigation methods applied between individual boreholes offer a solution in this direction. Active depth is different for embankment and for cutting. For the embankment, the depth depends mostly on serviceability limit state, with respect to the calculation of the embankment settlement. Boreholes or depth of investigation points should reach the depth where the stress increase is lower than roughly 20% of the initial vertical effective stress. A common recommendation is about 0.8–1.2 fold of the embankment height, at least 6 m. For cutting, the demanded depth is lower and depends on the ultimate limit state – slope stability. The depth should go about 1 m below the potential slip surface. A common recommendation is around 0.4 of the depth of cutting, but at least 2 m. Exceptions from these recommendations are cases when the subsoil has some cavities. The cavities can be either man-made, as the result of old mining, or natural, as in the case of karstic underground.

Presently, more attention is devoted to the cross section with respect to the width of the transport infrastructure strip. Slope stability and the recommended slope inclination are the most important aspects. Figure 3.3 illustrates slip surface possibilities for cuts, particularly

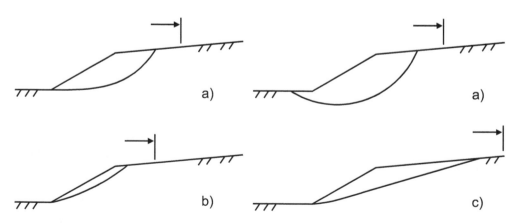

Figure 3.3 Demanded range of GI in cross section depending upon expected potential slip surface

when natural ground is inclined. Very dangerous is the case when natural ground is susceptible to sliding. Investigation points should help to solve the problem.

Nowadays there is a tendency to prevent the consumption of natural ground (greenfields).
Steeper slopes with application of slope reinforcement will be covered in Chapter 4. This
question is also sensitive due the complicacies with land buyout.

The question of whether to prioritize a gentle slope with higher consumption of land or a
steeper reinforced slope with lower land consumption should be solved at the early phases
of the ground investigation. The investor (owner) plays a very important role is this decision.

3.2.2   Geotechnical/Ground investigation report – GIR

The ground investigation report is the basic background material for the design and execution of the geotechnical structures. The basic aim of GIR is to provide the most reliable
information possible for the next follow-up phases of the structure design and construction.
Therefore, all steps of the GIR have to consider the character of the proposed geotechnical
structure.

With respect to *GI planning*, it is furthermore:

- The level of the previous information about locality – the results of the desk study;
- Grounds for the areal range of GI;
- Grounds for the depth range of GI;
- Grounds for the selection of the investigation methods;
- The statement that the proposed techniques are in agreement with standards and gen-
  eral recommendations. In some cases it also includes the specifications demanded by
  national investors – e.g. for construction of motorways and railways.

With respect to the *field and lab tests*:

- Grounds for the selection and the range of individual tests;
- The statement that all tests were performed and evaluated in agreement with existing
  standards (with the exception for nonstandard tests).

With respect to the *GI and tests results interpretation and evaluation*: This part is most
important and deserves detailed specification.

Accuracy of the ground model has also a narrow relationship to soil structure, namely to
macrostructure. According to Morgenstern (1969), the macrostructure can help in solving
engineering problems. Knowledge of detailed stratigraphy of the locality is a very important aspect of macrostructure study, especially as some sediments are created by variation
in small layers, more or less different from each other. Varved clays of glacier origin are a
typical case. Similarly, variation in grain-size distribution is observed for coarser material
settling – sands and gravels. This variation has an important impact on permeability, e.g.
for Danube river sandy gravel subsoil the permeability in the horizontal direction is roughly
100 times higher than in the vertical direction (Myslivec *et al.*, 1970).

Extended inspection of the drilling core can help to identify small layers, which can significantly influence the stability of the earth body – structure. Slim lamina with low shear
strength can control slope stability. A thin layer of organic composition has a strong impact
on deformation. A thin layer of sand influences – in a positive direction – the consolidation

of the surrounding clay material. Identifying such zones makes higher demands on the geo-technical investigation. The possibility of direct observation is an advantage. Also, the application of TV camera technique in narrow boreholes can help in this direction.

A similar effect can be observed with different discontinuities, cracks and zones of weakness. Over-consolidated fissured clays are a typical example. Fissures lower shear strength, namely when filled with softer clays. Fissure orientation is therefore also very important. As over-consolidated clays are sensitive to progressive failure, this type of failure will be covered separately. From the aforementioned, it is clear how important the fourth column is to geotechnical engineering (Figure 1.1).

Another sensitive problem consists with diversifying two layers with very close properties. The possibilities will be shown for the borrow pit. Whether the borrow pit is homogeneous or consists of more than one set of close soils determines different recommendations with respect to soil compaction and soil properties (after compaction). When statistical methods are applied, their predictive value depends on the number of tests. Most often, the evaluation in this direction is based on index properties of soils. Graphical evaluation of index properties is a typical case. Figure 3.4 shows the frequency of one variable, as is, e.g. a percentage of a certain soil fraction, moisture content, liquid limit etc.

By comparing two variables, some basic information about tested soil can be determined. Well known is the plasticity diagram (with variables w_L and I_p) which is used for fine soil classification (Figure 3.5). The triangular diagram, with three variables, is typical for soil classification based on percentage of soil particle fractions (g: gravel, s: sand and f: fine soil), see Figure 3.6.

Soil (rock) classification is the first step of soil characterization. The main aim of the classification system is to determine the name and symbol for a given soil. The ratio of the individual particle fractions (graphically illustrated in the triangular diagram) creates the base for the symbol and the name of coarse soils (with particles up to 60 mm). For fine soils, the plasticity diagram (Figure 3.5) creates the supplemental base for the name and symbol determination. Very coarse particles, cobbles (60–200 mm) and boulders ($\geq$ 200 mm) should be treated separately.

EN ISO 14688–1 and EN ISO 14688–2 are used now to create the main frame for soil description and classification. Similarly, EN ISO 14689 is used for rocks. However, there are many other classification systems. The Unified Soil Classification System (USCS) is

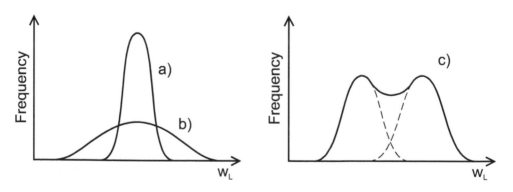

Figure 3.4 Frequency (different distribution) for one variable (w_L): (a) narrow band distribution; (b) wide band distribution; (c) distribution for two close sets

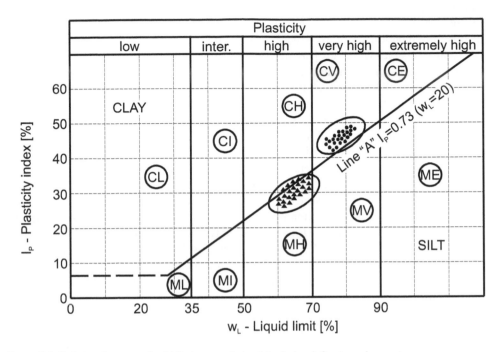

Figure 3.5 Relation between liquid limit w_L and plasticity index I_p for two close sets

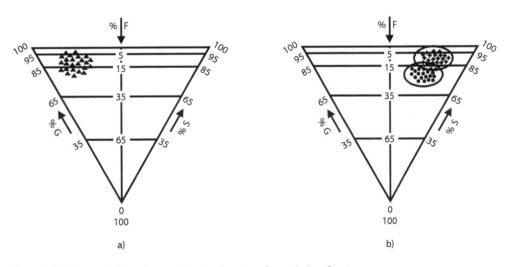

a)   b)

Figure 3.6 Three variables: three grain-size fractions for soil classification

probably the most widespread. A. Casagrande originally developed the system, primarily for new field airports subsoil determination during World War II. With small modifications, this system is used in many countries of the world.

Second main aim and very important advantage of the elder classification systems is a specification of natural soil properties. But in this case the soil state characterization is needed – consistency (with help of index consistency I_C) for fine (cohesive) soils and respective relative density (with the help of relative density index I_D) for coarse (granular) soils. Based on long-term experiences with individual types of soils, standard tables of representative/characteristic soil parameters were created. The standard tables are part of many textbooks on soil mechanics, or are part of national standards and codes (e.g. CSN 73 1001:1987). A significance of such tables is also highlighted by persons close to EC 7 (e.g. Bond, 2013). Soil classification together with geotechnical data estimation is specific for materials from the borrow pit which is proposed for earth structure construction. Classification during the evaluation of soil in the borrow pit, which will be used for earth structure construction, is based not only on index properties but also on an assumption that the soil is well compacted with moisture-density control. Suitability of soils for earth structures of transport engineering features (e.g. CSN 72 1002:1993) are supplemented by geotechnical data. Suitability of soils for earth structures of hydroengineering features (e.g. Sherard *et al.*, 1963) give an approximate correlation between embankment properties and soil classification groups (based also on USCS). However, nowadays CEN/TC 396 Earthworks, published set of standards EN 16907, when EN 16907–2 – Part 2 is devoted to classification.

Note: Type of soil can also be distinguished from some field tests. One of such first proposal for cone penetration tests presented (Begemann, 1965) on the basis of different results for cone resistance and for resistance measured along the sleeve, now denoted as q_c (MPa), respective f_s (MPa) (Lunne *et al.*, 1997).

The determination of mechanical-physical properties during either field or lab tests is typical for geotechnical structures associated with higher risk (GC 2 and GC 3). Strength properties (shear parameters), stiffness properties (deformation parameters) and hydraulic properties (parameters of hydraulic conductivity – permeability) are measured directly in the lab. During field tests, the individual parameters are derived from measured readings and therefore denoted as derived parameters.

The main differences between field and lab tests are:

- Conditions in the case of field tests are closer to the real conditions than for lab tests. Samples collected for the lab tests undergo stress changes and therefore in important cases should, during the first step, be reconsolidated in initial *in situ* stress conditions. For lab tests, the size of the tested sample plays an important role, namely for fissured clays, with a great impact on permeability and strength parameters.
- Lab tests allow sample testing under the conditions which can occur during geotechnical structure construction and structure operation. For the hydraulic properties, this comprises hydraulic gradient, degree of saturation and state of stress. For the stiffness properties, it is 1D or 3D loading, drained or undrained conditions and evaluation of deformation parameters for expected range of stresses. For important projects, stiffness for very small strains can be measured. Finally, for the strength properties, it is possible to simulate different design situations under drained or undrained conditions, pore pressure measurement during loading, all for expected range of stresses. For important projects, the stress-strain relationship for expected stress or strain paths can be evaluated.

The importance of detailed description of the individual tests is obvious from the aforementioned, not only that the tests were performed under conditions described in different standards, but also that they are more or less simulating the change of conditions of individual elements in the geotechnical structure, with a great predictive value for the designer.

To summarize the chapter, the main outputs of Ground/Geotechnical investigation report are:

- Ground model in graphical form, for earth structures of transport engineering, is longitudinal section supplemented by cross sections. Individual layers, reported also as a geotechnical type or unit, can contain standard geological age-based classification, namely in the case when rocky subsoil was reached. For soils (usually quaternary age), genesis (sediments or residual soils) is supplemented by first judgement based on grain sizes (clay, silt, sand, gravel, cobbles, boulders). However, the type of soil, based on the soil classification system, is most important. Depth of the groundwater table and its fluctuations are part of these sections.
- Geotechnical data, either index properties or mechanical-physical properties, are presented in table form. Where there is sufficient number of results for statistical evaluation, the histograms, illustrating the range of values and their distribution, are a very useful form of presentation. The same procedure is valid for the borrow pit.
- Personal evaluation of all phases of GI, preferably with respect to the interaction of designed geotechnical structure with ground. Whereas this evaluation is based on previous experiences with similar ground or geotechnical structures. Commentary to the obtained geotechnical data is also part of this evaluation, especially when the data fall outside of expected values.

3.3 Geotechnical design model – characteristic values determination

The specification of the geotechnical design model is a crucial step in the whole geotechnical design process. In principle it is coming out from all previous information, with the main aim to select for a given problem (geotechnical structure design) the geotechnical parameters which are competent for this problem. These parameters can be called representative values; EC 7 (2004) uses the term "characteristic values." At the end of the characteristic values selection process, the geotechnical design model is specified. Characteristic values are allocated to each lithological layer. For the simple geological model presented in Figure 3.2, the characteristic values can be attributed to the individual lithological layers, Figure 3.7. In agreement with the model indication, these values are subsequently used for the design in the calculation model.

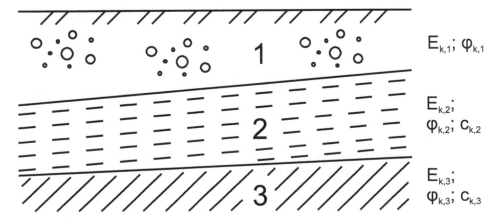

Figure 3.7 Geotechnical design model with characteristic values of the geotechnical parameters

Selecting the characteristic values of the geotechnical parameters is therefore a very important step and as a whole strongly influences the final design. Ideally, they should be very close to the real values, but a little bit on the safe side, namely for ULS.

In accordance with EC 7, the selection of characteristic values for geotechnical parameters shall be based on:

- The measured results and derived values from laboratory and field tests, complemented by well-established experience;
- A cautious estimate of the value affecting the occurrence of the limit state.

These requirements emphasize the significance of all previous information contained in the ground/geotechnical investigation report, together with an evaluation of all the limit states which can be relevant for a given design. Therefore, the designer, who is responsible for the safety of the design, has to check the sensitivity of the characteristic values for the selected limit state. This is another way of saying that the characteristic values for individual layers can be different when a different limit state is assessed. For the process of characteristic values selection, Frank *et al.* (2011) indicate that each word in the second requirement is very important:

- Selected – emphasizes the importance of engineering judgement.
- Cautious estimate – some conservatism is required.
- Limit state – the selected value must relate to the limit state.

At the same time, EC 7 indicates some other factors that a cautious estimate should take into account:

- Geological and other background information, such as data from previous projects;
- The variability of the measured property values and other relevant information, e.g. from existing knowledge;
- The extent of the field and laboratory investigation;
- The type and number of samples;
- The extent of the zone of ground governing the behaviour of the geotechnical structure at the limit state being considered;
- The ability of the geotechnical structure to transfer loads from weak to strong zones in the ground.

The last two points require a supplemental comment.

The significance of the extent of the zone of ground governing the behaviour of the geotechnical structure at the limit state can be shown with the example of two different embankments with the same height (guaranteeing the same contact pressure in the centre) but with different width. A wider embankment (which assumes a longer potential slip surface) is better able to balance a stronger zone with a weaker one along the longer slip surface than is a narrower embankment (Figure 3.8).

The ability of the geotechnical structure to transfer loads from weak to strong zones can be demonstrated for different distributions of weaker zones below embankments. When these weaker zones are closer to each other, this ability decreases (Figure 3.9). A similar case is typical for pile foundations. The undiscovered small weaker zone beneath a single pile can have a fatal impact, while for the group of piles the negative impact is much lower.

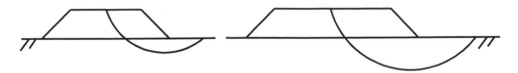

Figure 3.8 The significance of width of the embankment on the selection of characteristic values

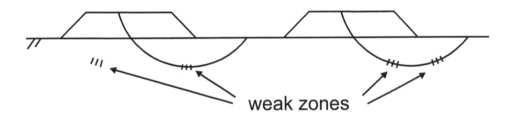

weak zones

Figure 3.9 Different negative impacts of weak zone beneath the embankment

As mentioned previously, there are two basic approaches to the selection of the charac-teristic values of the geotechnical parameters. The EC 7:2004 specifies the conditions for application of these two basic approaches:

• Standard tables approach (2.4.5.2 (12)) – "when using standard tables of characteristic values related to soil investigation parameters, the characteristic value shall be selected as a very cautious value."
• Statistical approach (2.4.5.2 (10)) – "if statistical methods are employed in the selec-tion of characteristic values for ground properties, such methods should differentiate between local and regional sampling and should allow the use of a priori knowledge of comparable ground properties."

As selecting between these two approaches is a very sensitive issue, the following sec-tions are devoted to them.

3.3.1   Standard tables approach

Standard tables are the result of long-term experiences; they are the expertise of our prede-cessors. Basically, these standard tables express the relationship between index properties and mechanical-physical properties, either for the ground in its natural state or for the com-pacted ground used for earth structures. The significance of this experience is expressed also in the statistical approach – as the selection of characteristic values should also be based on "a priori knowledge of comparable ground properties."

The decision of whether this approach is sufficient for the designed geotechnical structure should be solved in the early stages of the whole process. EC 7 does not specify directly in which cases standard tables approach can be used, but this indirectly results from the risk with which the design is associated. It is evident that only index properties of ground will be sufficient for structures with low risk, with low consequence of failure. The decision in this direction makes it possible to select a corresponding sampling category.

Standard tables utilize different classification systems. For particular soil, after the name and type specification, supplemented by state characteristics (consistency for fine soils and relative density for coarse soils), the range of expected mechanical-physical properties such as deformation properties, strength properties or filtration properties are given in these tables. From this range, the characteristic values should be selected with great caution, taking into account all aspects already mentioned.

The majority of standard tables are from the Unified Soil Classification System or from different small national modifications, nowadays also from soil classification used in EN 16907–2. EN ISO 14688 and 14689 utilize a slightly different classification, however up to now without any recommended values of geotechnical parameters.

For the Czech Republic, a small modification was proposed by Vaníček (1987), which was consequently used for most of the national geotechnical codes, not only for natural ground (CSN 731001:1988) but also for fills of small dams or roads. A more detailed description of names and symbols was presented by Šimek et al. (1990) or by Vaníček and Vaníček (2008). Practical application of this classification together with standard tables is described by Vaníček and Vaníček (2013b) for the design of spread foundations.

Nevertheless, it is necessary to stress that such tables do not need to have general validity, as the correlation used for these tables was sometimes discovered for a certain geological environment as well as for a certain type of geotechnical structure. Therefore, they can be used only with great precaution.

3.3.2   Statistical approach

Statistical evaluation of the results from the lab or field tests requires a certain minimal number of results for an individual lithological layer – geotechnical type. Usually five results are denoted as sufficient (if the results distribution can be rated as a statistical set). This demanded number, especially for laboratory tests performed on samples based on sampling category A, is relatively strict and is typical for geotechnical structures with high risk – for Geotechnical category 3. A certain advantage of the field tests, namely for the penetration tests, is easier fulfilment of the condition connected with a minimal number of results.

Nevertheless, the application of statistical methods is always a difficult task, which we now focus on. For example, Orr et al. (2002) indicate on the basis of evaluating the practice in different countries, that most of them prioritize selecting the characteristic value as mean value ± standard deviation/2, or $X_k = X_{mean} \pm SD/2$.

For man-made materials, the characteristic value is usually selected as worst credible value governing the occurrence of the limit state under consideration – the characteristic value is often defined as a (5%) fractile; the characteristic value should be derived such that the calculated probability of a worst credible value governing the occurrence of the limit state under consideration is not greater than 5%. This assumption for geotechnical structures can lead in many cases to a great conservatism and can be accepted only for cases where the zone of influence is small. For a large zone of influence, it is highly likely that the real value will be somewhere between this minimum value and a mean value, with a probability of 95% that the mean value governing the occurrence of a limit state in the ground is more favourable than the characteristic value.

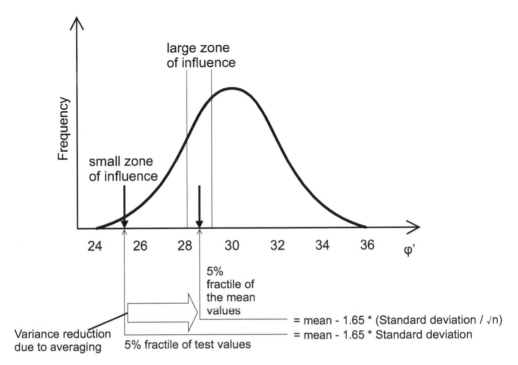

Figure 3.10 Recommended specification of the range from which the characteristic value should be selected

Source: According to Schneider.

One possible way to solve this problem was proposed by H. Schneider during the CEN250/SC7/WG1 meeting in 2014. For the distribution curve of the angle of friction, specified are the aforementioned two values, where the lower one represents a 5% fractile of the test values and the second one the 5% fractile of the mean values (Figure 3.10).

The variance between these values depends on the size of the zone influenced by the proposed geotechnical structure.

The first generation of Eurocode 7 gives relatively great freedom to the individual approach of selecting characteristic values, when using the term "cautious estimate." The second generation of Eurocode 7 (prEN EN 1997–1:2018), which is now in the phase of elaboration, will be more specific. For the characteristic value X_k evaluation, the following equation is proposed:

$$X_k = X_{av}[1 \mp k_n \Delta_x] \tag{3.3}$$

Where:

X_{av} is an estimate of the average value of the ground property;

Δ_x is an estimate of uncertainty affecting the ground property;

k_n is a coefficient that depends on the number (n) of site-specific data used to estimate X_{av};

$\mp$ denotes that $k_n \Delta_x$ should be subtracted when a lower value of X_k is critical and added when an upper value is critical.

The main difference for the various geotechnical categories consists in specifying the average value X_{av}. For GC 1, this average value can be based solely on a so-called pre-assessed value, when the indicative average values of properties for some ground types will be given in table form. For GC 3, this average value will be based only on measured values, while for GC 2 the combination of both values will be typical. A detailed procedure will be described for individual geotechnical structures. The national standards bodies will have a great freedom, especially to present and use their own indicative (table) values.

From the aforementioned, it is obvious that the selection of characteristic values can be performed only during the phase of the geotechnical structure design, not before, e.g. during ground investigation (e.g. Bond and Harris, 2008). This statement is in agreement with the responsibility of the designer for safe design. Nevertheless, the cooperation of all partners is the basic premise.

3.3.3 Characteristic value for improved ground

Earth structures differ from other geotechnical structures as foundation structures or under-ground structures, as in the case of earth structures we have to consider the two basic types of ground: natural and man-made. For the other two geotechnical structure types, we have to determine only the properties of the natural geological environment.

However, for both basic cases of ground, subsoil and fill, ground improvement methods can be applied. For fill, not only natural aggregates can be used but also alternative aggregates. From the viewpoint of determining characteristic values, the methods of ground improvement can be divided into two subgroups:

- Diffusive ground improvement;
- Discrete ground improvement.

For diffusive ground improvement, the design has to consider new properties for the geological environment as a whole, or more precisely stated as for individual quasi-homogeneous layers of this environment. For discrete ground improvement, soil parameters keep the original values, and parameters for inclusions have to be specified and the limit state of failure for inclusion should be controlled.

Different ground for earth structures

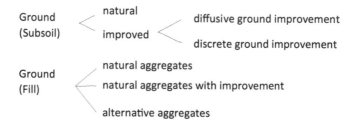

The ground improvement is generally executed after the earth structure is designed. During the design process, the assumption is made regarding the values of improved ground properties. Improvement is the result of the pilot study during which different alternatives of modification for natural ground are performed and tested in order to meet the design assumptions.

For the discrete ground improvement, the determination of its properties is an easier problem, as properties of the ground itself are generally not as affected by this ground improvement. Therefore, attention is concentrated on the characteristic values of this discrete element. For man-made materials, as e.g. steel nails applied for slope of cuts improvement, it is again not a great problem, as the producer of this material can guarantee its parameters. For vertical discrete elements – different types of piles, lime columns, geotextile encased columns etc. – the designer can count on previous experience to specify the characteristic values of this discrete element, as the limit state of failure of this element should be controlled during the design as well. Nevertheless, some improvement around these columns, for example lime columns, were observed, mainly due to reducing the water content and partly by the hardening of an annular zone around the lime columns (e.g. Holeyman and Mitchel, 1983). A short note can be added to this section. Vertical discrete elements are mostly applied to improve the subsoil of the transport infrastructure fill, and great care should be devoted to the contact of this fill with improved subsoil. In most cases, some horizontal platform is needed.

For the diffusive ground improvement, the properties of subsoil are modified, and the designer will have to select how much the improvement is needed. Most methods are based on the principle of speeding up subsoil consolidation. Two methods are preferred for this – preloading and applying vertical drains. The first one is applied for soft subsoil, and during the process of preloading the porosity is significantly decreased and soil parameters improved. The second method reduces the pore pressure in the subsoil during embankment construction and in this way increases shear strength and the speeding up of vertical deformation. Nevertheless, the designer counts with prognosis on which the design of improvement is based and the reality can be controlled only by monitoring both settlement and pore pressure dissipation as function of time. Therefore, the observation method of the design can be very useful in this case, as the speed of embankment construction can be corrected.

Application of the improved fill is typical for the cases where some less appropriate material should be used, the main aim is to decrease total volume of the fill (to safe natural aggregates) or steeper slopes of fill are proposed (to decrease the consumption of land).

The geosynthetics application is a typical example for a fill discrete ground improvement. Especially geosynthetics with high tensile strength can significantly improve the behaviour of a fill as a complex. Soil reinforcement with geosynthetics is now used often, and the main attention devoted to determining the short-term and long-term properties of this reinforcing element and their participation in load transfer. More specification will be given in Chapter 4.

Soil stabilization is on the other side a typical method for diffusive ground improvement. With respect to the earth structures, attention is devoted to improving the soils in layers, which are compacted. Natural aggregates are mixed with some additives, typically with lime, cement or gypsum. The advantage is that this mixture can be prepared in the laboratory in advance, so that the properties can be determined in the investigation and design processes. More specification will be discussed in Chapter 4.

Fill from alternative aggregates is a great chance for natural aggregates savings. The amount of waste or different by-products has still growing tendency. Great effort is now devoted to solving this problem, as we are obliged to the next generations to do something about it, to start behaving more sustainably. From the view of different waste or by-product application in earth structures of transport engineering, such alternative aggregates are preferred. Examples include fly ash, waste rock, slag, construction and demolition waste or soil/rock excavated during the construction of underground structures. This will also be discussed in greater detail in Chapter 4.

Finally, determining the characteristic values of geotechnical data for improved ground and fill is even more difficult. The process of data determination is particularly sensitive for improved ground (subsoil), as there is limited possibility to control it. In many cases, the contact between fill and ground needs improvement, as does the subsoil reinforcement with vertical elements. These problems will be discussed later, as they are not strongly connected with characteristic values determination.

3.3.4   Characteristic values of groundwater pressure

Today opinion prevails that the characteristic values of groundwater pressure can be connected to maximum or minimum levels (according to what poses the most danger to a respective structure) and to what is expected during the structure's life expectancy (or for some other time margins). For free water, e.g. during floods, the maximum level can correspond to the flood level with a period of 1.5 times higher than lifetime expectancy. For the groundwater level or piezometric level, the same is valid. The advantage of this approach is the fact that there is no need to apply a partial safety factor on loading, γ_F.

The problem is now in the centre of interest, namely concerning climate changes. The prognoses are generally based on historical data, and the question is how much this data will be valid in the future.

3.4   Calculation model

The calculation model can and must be different for different structures and different geological (geotechnical) conditions. Only simple structures, classified as GC 1, can be designed on experiences gained up to now. For all other structures, individual limit states should be checked.

3.4.1   Limit states division

As already mentioned, two basic limit states have to be checked:

- ULS – ultimate limit state
- SLS – serviceability limit state

The second one involves deformation control, whether the calculated value is lower than the acceptable value.

Ultimate limit state involves structure failure check and considers the possibility of:

- Failure of ground – what the most typical case is for earth structures;
- Failure of structural element in contact with the ground – what a typical case is for structural elements used for ground or fill improvement (e.g. failure of soil nails or geosynthetics);
- Combined failure of the ground and structural elements – again, what a typical case is for ground improvement – in this case, however, the plane of failure passes through both the ground and the reinforcing element.

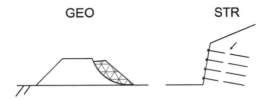

Figure 3.11 Different types of ultimate limit states for earth structures based on the first generation of EC7 – types STR and GEO

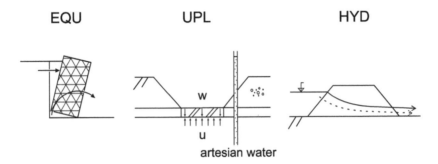

Figure 3.12 ULS – types EQU, UPL and HYD

The character (mode) of failure can be different, and particularly for the earth structures the different cases are illustrated in Figures 3.11 and 3.12:

- Failure along plane of failure as is a case of slope stability or bearing capacity of subsoil below embankment, where shear strength (in limited cases also tensile strength) plays the most important part. EN 1997 denotes this limit state as GEO.
- Failure due to internal or surface erosion – when hydraulic gradient creates soil particle movement. EN 1997 denotes this limit state as HYD.
- Failure due to uplift – when uplift static water pressure is higher than geostatic pressure. EN 1997 denotes this limit state as UPL.
- Failure due to loss of static equilibrium. EN 1997 denotes this limit state as EQU.
- Failure of the structural element (STR).

The first generation of EN 1997 denotes the different modes of failure as previously described, but the second generation drops such description.

3.4.2 Analytical and numerical calculation models

A calculation model, according to general practice and Eurocode 7, may consist of any of the following:

- An analytical model;
- A semi-empirical model;
- A numerical model.

Our attention will be focused on analytical and numerical models, as they are preferred for earth structures of transport engineering. Basic differences were mentioned previously – analytical models solve ULS and SLS separately, while numerical models can solve them in one step. Analytical methods are covered in classical books on soil mechanics and numerical methods in special publications such as Zienkiewicz and Taylor (2000) and Potts and Zdravkovic (1999, 2001). Input data for these two models are also different, and those conducting lab and/or field testing should know in advance what input data are needed. For SLS and ULS of the type GEO, the analytical models require strength parameters (shear strength, respective tensile strength) and deformation parameters (for both loading and unloading). For the same case, the numerical models need stress-strain characteristics, again for loading and unloading, respective stress-strain characteristics for tensile zone (Vaníček, 2013c). Basic stress-strain characteristics for loading are presented in Figure 3.13, and we are speaking about constitutive relationships or constitutive models.

The authors described in previous publications detailed specifications of different constitutive models, e.g. Vaníček and Vaníček (2008). Briefly:

- The first graph (a) assumes a linear relationship between stress and strain – it represents the linear elastic model.
- The second graph (b) shows only plastic deformations – it represents an ideal plastic model.
- The third graph (c) represents the linear elasto-plastic model. This alternative is reasonably close to the behaviour of cohesive, normally consolidated soil or loose sand, whose working diagram is shown in graph (d).
- The model of the real cohesive over-consolidated soil or dense coarse soil is shown in graph (e) and can be approximated by linear elastic ideal plastic model with softening that is shown in graph (f).

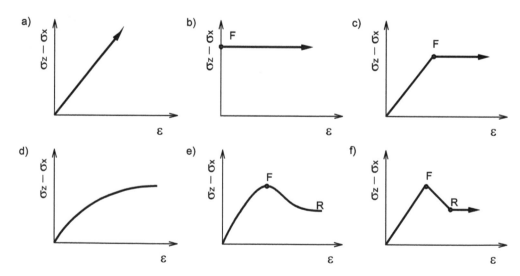

Figure 3.13 Various constitutive models

The behaviour of normally consolidated soil (d) is therefore possible to model via bilinear relationship (c) or by hyperbolic relationship proposed by Kondner (1963):

$$(\sigma_z - \sigma_x) = \varepsilon_z/a + b.\ \varepsilon_z \tag{3.4}$$

The alternative is to divide the relationship into several steps, where for each given stress level are defined pseudo-elastic parameters derived from the real relationship between stress and strain, using either tangent moduli (Clough and Woodward, 1967) or secant moduli (Girijivallanhan and Reese, 1968), Figure 3.14.

Basic differences between outputs from both models – analytical and numerical – were mentioned already by Dunlop and Duncan (1970), namely for the case of slope stability for cut, Figure 3.15.

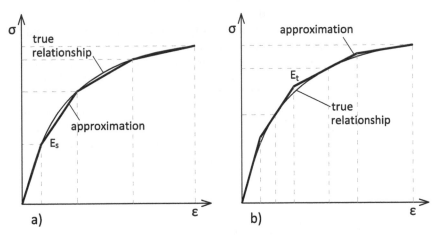

Figure 3.14 Possibilities of substitution of real behaviour by secant moduli or by tangent moduli for the given stress level

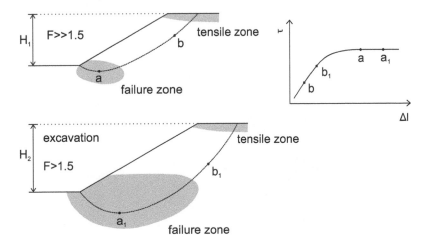

Figure 3.15 Various interpretations of the analytical and numerical models for cut stability

For the numerical model (FEM), they used the bilinear relationship between stress and strain, and once the peak strength is reached, it becomes constant. The areas of plastic failure are enlarged with the continuous enlargement of the pit. Areas with tensile stresses are shown at the top part of the slope. Even when the plastic zones exist, the overall factor of stability F is still higher than one (calculated by analytical method using limit equilibrium approach along circular slip). The reason is that in the elastic zone, there is still a large reserve in soil strength.

Usually the numerical calculation model is more demanding on one side, but on the other one is also giving very important results about displacement of individual elements. When applying the principle of the partial factors of safety according to the Eurocode 7, however, the solution of ULS and SLS needs two different models, as for SLS the characteristic values of the geotechnical parameters are needed, while for ULS the design values are used – so it means that the characteristic values are transferred to the design values by the partial factor of safety on material γ_M.

Closer specifications about the numerical calculation model will be provided in the second generation of EC 7 (roughly in 2022–2023).

Therefore, to better describe the demands on two basic limit states, ULS and SLS, the main focus is now on the analytical calculation model.

From the different ultimate limit states, the limit state of GEO is most important for the earth structures for transport engineering. This limit state is associated with slope stability, as the strength of soil (or rock) is significant for providing resistance.

In accordance with the limit state principle, which is part of Eurocode 7, equilibrium on the slip surface is needed even if material properties are reduced by a partial factor of safety for material γ_M and the loading is increased by partial factor γ_F. Safe design of slope stability should fulfil the following condition (Figure 3.16):

$$E_d \leq R_d \tag{3.5}$$

Where:
E_d is the design value of the effect of actions;
R_d is the design value of the resistance to an action.

Note: By classical approach, the total factor of stability, marked as F or FoS (factor of safety), has guaranteed stability when $FoS \geq 1.3–1.5$. For a limit state approach, the stability is satisfied when:

$$R_d/E_d \geq 1 \tag{3.6}$$

and this ratio can be marked as FoS_d – design factor of safety, or F_{geo}.

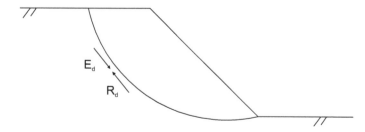

Figure 3.16 Equilibrium on potential slip surface (GEO)

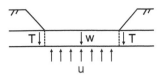

Figure 3.17 Conditions for failure by uplift (UPL)

Bond and Harris (2008) also recommend using the reverse ratio:

$$E_d/R_d \times 100\% \tag{3.7}$$

denoted as Λ_{GEO} – "degree of utilization" – which expresses the maximum resistance utilized for the demanded equilibrium (what reserve is still there).

The other ULSs, like UPL, EQU or HYD, which are usually associated with some special condition, e.g. the condition for the UPL limit state – loss of equilibrium of the ground due to uplift by water pressure, Figure 3.17, can be defined as:

$$U \leq W + T \tag{3.8}$$

Where:

U is total uplift water pressure (on an area A);

W is overall soil gravity which acts vertically downwards (on an area A);

T is potential friction resistance along the perimeter of the body susceptible to the uplift (with an area A).

When applying Eurocode 7 principles, partial factors are applied on each force. However, for uplift, water pressure is another possibility – to apply maximal expected artesian water pressure.

A typical case of the limit state of the EQU type is connected with loss of equilibrium of the structure (e.g. concrete retaining wall) or of the ground (e.g. geosynthetic-reinforced retaining wall), which are considered as a rigid body. In this case, the strength of the structural element or the ground is insignificant in providing resistance. For both aforementioned cases, the loss of equilibrium is controlled via wall overturning around the point A, Figure 3.18. Here, the basic condition can be defined as:

$$M_a \leq M_r \tag{3.9}$$

Where:

M_a is moment causing overturning: $M_a = E_a \times r_a$;

M_r is moment of resistance, consisting from $M_r = W_w \times r_w + E_p \times r_p$.

Where W_w, E_p respective E_a are wall weight, passive respective active earth pressures. Again, partial factors should be applied on them, fulfilling the Eurocode 7 design approach.

The limit state of the HYD type is typical for embankments, which are in contact with water. This case will also be discussed in Chapter 5 and is typical for the contact of

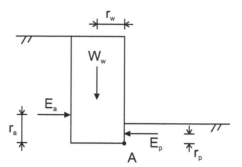

Figure 3.18 Failure by loss of equilibrium (EQU)

embankment with water during floods. Briefly speaking, failures are the result of seeping water, caused by hydraulic gradient *i*. Failures are denoted as:

- *Failure by heave* – when upwards seepage forces act against the weight of soil;
- *Failure by internal erosion* – when hydraulic gradient is causing the transport of soil particles within a soil stratum or on the contact of two different layers (one of them is a filter);
- *Failure by piping* – when erosion begins at the downstream face of embankment and regresses until a pipe-shaped discharge tunnel is formed.

Failure by heave can occur when water flow upwards. This situation can occur when there is a thin permeable layer below the small dam or dyke (surfaces of which are used for transport infrastructure). Upward water pressure can initiate failure by heave at the weakest point (Figure 3.19). The solution of the soil element equilibrium may be carried out both

- in total stresses:

$$u_{crit} \leq \gamma_d \tag{3.10}$$

when u_{crit} is the critical pore pressure and γ_d design value of total weight density

- in effective stresses:

$$p_{v,crit} \leq \gamma_{sub,d} \tag{3.11}$$

when $p_{v,crit}$ is the critical seepage pressure ($p_{v,crit} = i_{crit} \times \gamma_w$) and $\gamma_{sub,d}$ is a design value of effective weight density.

Failure by internal erosion can be controlled via:

- Condition $i_d \leq i_{crit}$;
- Fulfilment of filtration criteria.

Note: The main aim of a designed filter is to allow seepage but not particle movement. With time, different filtration criteria were proposed, either for grain filters of for geotextile

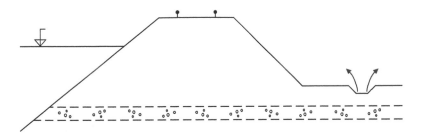

Figure 3.19 Failure caused by hydraulic gradient (HYD), conditions for failure by heave

filters (e.g. Terzaghi and Peck, 1948; Sherard *et al.*, 1963, 1984; Vaughan, 1976; Vaughan and Soares, 1982; Vaníček and Vaníček, 2008; ICOLD Bulletin 164 on internal erosion, 2017).

The main approach to prevent failure by piping leads via control of flow net – water cannot outflow on the downstream side as far as this face is not protected by filter.

In addition, different protective measures are used for the protection of the embankment against surface erosion, either for the erosion of the upstream face caused by waves, or for the erosion of the crest and the downstream face caused by embankment overflowing.

3.4.3   Ultimate limit state – GEO type – slope stability

Up to this point, many different slope stability methods were defined. The most widely accepted is Bishop's method. Therefore, the application of partial factors will be shown on this method when limit state of failure is solved in accordance with the EC 7. The basic version of Bishop's method is counting with circular slip surface typical for homogeneous cohesive soils and with division of sliding mass (annulus) on individual slices. For each slice next to moment equilibrium, the balance of horizontal and vertical forces is needed (see Figure 3.20). The final equation for total factor of safety for steady flow follows:

$$F = \frac{1}{\sum W \sin \alpha} \sum \frac{c_{ef} b + (W - ub)\, tg\phi_{ef}}{\cos \alpha + \dfrac{tg\phi_{ef} \cdot \sin \alpha}{F}} \tag{3.12}$$

When applying partial factors of safety according to EC 7, the equation has the different expression:

$$F_{geo} = \frac{1}{\sum W_i \sin \alpha_i} \sum \frac{c_d.b_i + (W_i - u_i b_i).\tan \phi_d}{\cos \alpha_i + \dfrac{\tan \phi_d.\sin \alpha_i}{F_{geo}}} \tag{3.13}$$

Respectively, when "degree of utilization" Λ_{GEO} is used, after that:

$$\Lambda_{geo} = \frac{\sum (W_{d,i} + Q_{d,i}) \sin \alpha_i}{\sum \dfrac{(c_{d,i} b_i (W_{d,i} + Q_{d,i} - u_{d,i} b_i)\tan \phi_{d,i})\sec \alpha_i}{1 + \tan \alpha_i \tan \phi_{d,i} \Lambda_{geo}}} \tag{3.14}$$

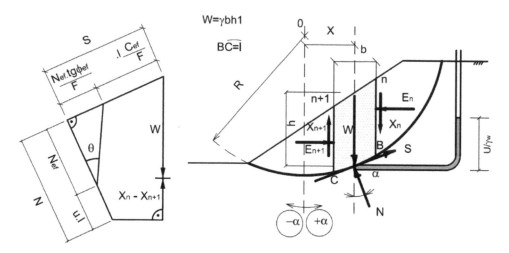

Figure 3.20 Forces acting on individual slice

3.4.3.1   Short- and long-term slope stability for embankment – fill

There are two basic cases with respect to the geological profile. For the first one, the better properties of subsoil are assumed, for the second one the opposite is true.

For the first case, the potential slip surface is passing first through the fill. In the case of short-term stability, after finishing the fill, undrained shear parameters can be used as an estimation of pore pressure development in the fill is complicated. As the degree of saturation after compaction is close to 1, undrained shear strength $\tau_u = c_u$ and is independent of confined pressure. The basic shape of Bishop's method equation is therefore reduces to:

$$F_{geo} = \frac{1}{\sum(W_i \sin\alpha_i)} \sum \frac{c_{ud}.b}{\cos\alpha} \tag{3.15}$$

For the first case and long-term stability, the excess of pore pressure had the chance to dissipate with time. Stability is improving with time, and the solution can be performed under a drained condition, with drained shear parameters φ' and c' with $\Delta u \div 0$, and pore pressure u is determined from the flow net for steady flow conditions.

For the second case, when the subsoil is very often called "soft" or weak, the potential slip surface is going through this subsoil, Figure 3.21.

Also, in this case the undrained solution can be applied as in aforementioned case. In this case, however, the solution with effective shear parameters, together with estimated pore pressures, is useful as well. This approach makes it possible to define critical pore pressure u_{crit} under which slope stability will be satisfied.

For the change of pore pressures Δu as a result of a change of main stresses $\Delta\sigma_1$ and $\Delta\sigma_3$, Skempton and Bishop (1954) recommended the relationship:

$$\Delta u = B (\Delta\sigma_3 + A (\Delta\sigma_1 - \Delta\sigma_3)) \tag{3.16}$$

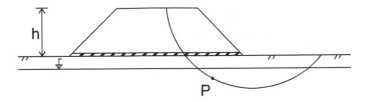

Figure 3.21 Typical character of slip surface for soft ground

Where A and B are coefficients of pore pressure. For the case of fully saturated soils, the coefficient $B = 1$ and is true for:

$$\Delta u = \Delta\sigma_3 + A\left(\Delta\sigma_1 - \Delta\sigma_3\right) \tag{3.17}$$

As the degree of soil saturation after the compaction is close to 1, the last equation can be applied. In this case, $u = u_0 + \Delta u$ is used in the basic Bishop's equation.

At the end of filling, the excess pore pressure reaches maximum value (Bishop and Bjerrum, 1960). With time due to dissipation, the value of pore pressure is decreasing, reaching a value corresponding to the groundwater stable condition, which usually equals the initial pore pressure u_0. A basic conclusion is the statement that slope stability is critical (lowest) at the end of filling; from the long-term stability point of view, the situation is significantly better. Determining pore pressure for long-term stability is much easier, and only drained shear strength parameters can be applied.

The methods of subsoil improvement leading to the excess of pore pressure decrease at the end of filling, often called methods of speeding up of consolidation, are applied when needed, when pore pressure exceeds the critical value. These methods will be described in section 3.4.5.1

3.4.3.2 Short- and long-term slope stability for cuts

For the case of cuts, the situation is reversed, see Figure 3.23.

Due to unloading, the pore pressure is decreasing, with a positive effect on slope stability. However, with time the pore pressure is increasing, reaching finally the level for steady flow. Therefore, the slope stability at the end of construction does not mean that slope failure will not occur with some delay. The investor (owner) should be very cautious, as the failure can occur beyond the guarantee period. On the other hand, this fact can be used in a positive way for the temporary slopes of the cuts.

The long-term stability calculation is performed with drained shear strength parameters.

For short-term stability, both approaches (with undrained or drained shear strength parameters) may be applied, first with the help of undrained shear strength parameters, namely with undrained strength c_u for saturated ground. More complicated is the solution for partly saturated ground, as there the undrained strength strongly depends on confining pressure. The solution with the help of drained shear strength parameters is also not as easy due to complicacy associated with prognosis of pore pressure changes.

The basic approach of how to decrease the probability of the cut failure with some delay is not to allow such increase of groundwater level. The application of different drains as horizontal drainage boreholes or trench drains are methods most often used.

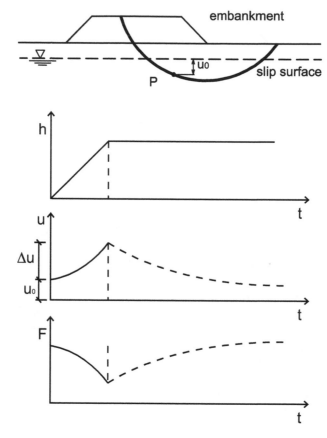

Figure 3.22 Changes of pore pressures with time at point P from Fig. 3.21 and the impact on stability

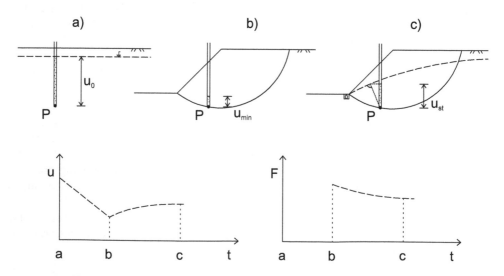

Figure 3.23 Changes of pore pressures (and stability) with time for cut

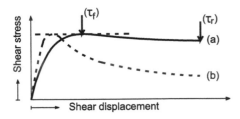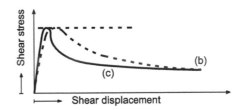

Figure 3.24 Typical drop-off of shear resistance with shear strain for over-consolidated clays

3.4.3.3   Slope stability for cuts in over-consolidated clays

Slope stability for cuts in over-consolidated clays deserves special attention. Shear resistance for such materials after reaching peak value τ_f is decreasing up to the residual shear resistance τ_r, Figure 3.24. The different character of this decrease is connected with the so-called index of brittleness according to Bishop (1967):

$$I_B = (\tau_f - \tau_r)/\tau_f \qquad (3.18)$$

As the shear displacement is highest at the slope toe, the progressive failure starts close to this slope toe, as there shear strength first is reaching peak value followed by a strength decrease up to the residual value. So it means that the failure is spreading progressively. A very low index of brittleness presents high sensitivity to progressive failure. However, for shear resistance development with an equal index of brittleness, there can be also differences, as is obvious when curves ad (b) and ad (c) are compared, as the last one is more critical.

Bjerrum (1967) states three conditions of a progressive failure:

1. The soil shows brittle behaviour and a striking lowering of strength at the deformations greater than that at which the peak strength is reached – a basic quality of strongly over-consolidated clays.
2. There exist places of stress concentration – the simplest is with fissured clays at corners of micro-fissures where the strength can be exceeded locally and so a progressive failure can start.
3. Limit conditions are given by different deformations – for example by a stress condition at the toe of the excavation where it causes various deformations, which, in many cases, exceed the deformation necessary for reaching the peak strength.

A proposal for which shear strength parameters can be used in this case is offered by Chandler and Skempton (1974), who discovered that for two types of over-consolidated clays, the failure happens at the peak angle φ', but practically at zero cohesion. The discussion in this way is still going on, particularly with the critical state soil mechanics approach (Schofield and Wroth, 1968; (Atkinson and Bransby, 1981; Bolton, 1979; Wood, 1990, all from the Cambridge University). They are working with a critical state angle of friction $\varphi'cs$ and with zero cohesion.

3.4.3.4   Slope stability along previous slip surface

Only in this case the residual shear parameters can be applied. Therefore, the identification of old slip surface already during the phase of geotechnical investigation is so important. However, first signal about this possibility can be obtained from engineering geological maps of landslide prone areas. The Czech Geological Survey is the main source of such information for the Czech Republic.

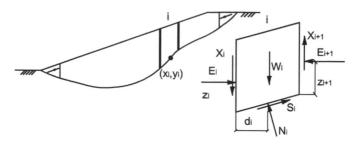

Figure 3.25 Conditions of strip equilibrium for general character of slip surface

The large shallow plane landslide, which affected the construction of motorway D8 between Prague and Dresden in 2013, is an example of landslide, which occurred along previous slip surface, Figure 5.13 (Jirásko *et al.*, 2017). Details will be provided in Chapter 5.

3.4.3.5   Slope stability for general slip surface

The previously discussed circular slip surface is the best fit for homogeneous soils. However, very often soil is heterogeneous, with significant bedding and with different zones, as for sandwich fill etc. For these cases, the slip surface character is irregular, non-circular, general.

In principle, the stability calculations for general slip surfaces can be divided into two groups:

* Method of slices creates a first group, where similarly as for circular slip surface the slope above the slip surface is divided into slices; however, the slip surface can be a general one, composed from individual line segments or from individual curvature parts, see Figure 3.25.
* Wedge methods create a second group, which solve the stability of a specific block above the selected slip surface, which is loaded from both sides by active and passive blocks of soil – but this solution is not typical for earth structures of transport engineering.

METHOD OF SLICES

The following methods belong into this group, for example Janbu (1973), Morgenstern and Price (1965) and Sarma (1973). Under certain conditions, the conventional method of Skempton and Hutchinson (1969) can be used as well. Some problems of these methods are given by redundancy (static uncertainty). Under the assumption that the slope above the slip surface is divided into n slices, the total number of unknowns is $6n - 2$ and the number of possible equations is $4n$. Detailed analysis of this fact is given, for example, by Sarma (1979).

Authors of slope stability methods of slices are introducing supplemental criteria and assumptions in order to obtain an acceptable solution with minimum effort. Therefore, the specification of these supplementary criteria and assumptions for the selected method in the GDR (Geotechnical design report) is critical for the subsequent design check.

3.4.4   Serviceability limit state

The serviceability limit state is very important for earth structures of transport engineering. These limit states are associated primarily with the deformation of subsoil, with fill or just with the upheaval of bottom of the pit excavation, see Figure 3.26. However, most important is the result of these deformations on the earth structure surface, either in longitudinal or cross sections. For different transport infrastructure as motorways, railways or airport

a) uncompressible

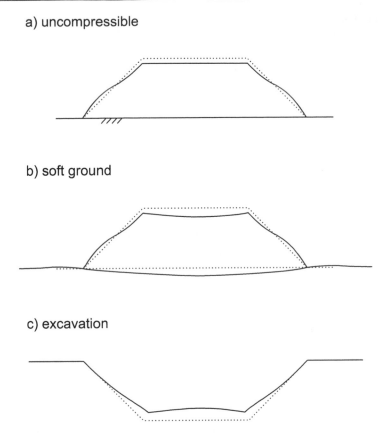

b) soft ground

c) excavation

Figure 3.26 Basic examples of earth structure deformations

runways, there are different demands. Usually stricter conditions are connected with the higher speed of means of transport.

Also in this case, it is necessary to distinguish between short- and long-term conditions – between settlement at the end of construction and settlement after some years when the consolidation of ground is finished. Short-term conditions are generally valid when the structure is being taken over by the investor (owner) from the construction firm. The demanded conditions have to be fulfilled at this time. Nevertheless, there is a second question – are these demanded conditions also fulfilled at the end of consolidation? Therefore, in many cases some methods for speeding up the consolidation process are designed and realized.

As the deformation is proportional to the stress changes and inversely proportional to the deformation characteristics, two basic conditions have to be fulfilled:

• The manner of determining the initial state of stress together with determining stress changes – this process is usually very well described in soil mechanics textbooks or in different standards specification.
• The manner of determining deformation characteristics, again with agreement with standards – CEN/TC 341 "Geotechnical Investigation and Testing" created the set of EN ISO codes.

Deformation characteristics should be determined for the expected range of stress changes. For more specific problems, the testing for expected stress paths can be very useful. However, the most often used standard deformation characteristics are specified in Table 3.2.

From this table, it is obvious that the oedometric modulus of deformation E_{oed} can be used only for final (total) settlement (s_c) under 1D conditions. For 3D conditions:

- Short-term settlement (s_i) uses the modulus of deformation under undrained conditions E_u and Poisson's ratio for undrained conditions v_u (which for $S_r = 1$ is practically equal to 0.5).
- Long-term settlement (s_c) uses the modulus of deformation under drained conditions E_{def} and Poisson's ratio for drained conditions v_{ef}.

Standard types of soil element deformations, either for 1D or 3D conditions, are shown in Figure 3.27 (e.g. Vaníček and Vaníček, 2008).

Note: For 1D deformation, other deformation characteristics expressing the relation $\Delta\varepsilon = f(\Delta\sigma')$ and/or $\Delta e = f(\Delta\sigma')$ can also be used where e is the void ratio. Among the most commonly used deformation characteristics are:

- Coefficient of volume compressibility: $m_v = \Delta\varepsilon/\Delta\sigma' = 1/E_{oed}$
- Coefficient of compressibility: $C = 1/\Delta\varepsilon \times \Delta\ln s'$
- Compressibility ratio: $a_v = -\Delta e/\Delta\sigma'$
- Compressibility index: $C_c = -\Delta e/\Delta\log s'$

An advantage of deformation characteristics based on the change of the void ratio e is some control over porosity during the whole loading process.

Table 3.2 Basic deformation characteristics for 1D and 3D deformation and for undrained and drained conditions

	Undrained conditions	Drained conditions
1D	–	E_{oed}
3D	$E_u; v_u$	$E_{def}; v_{ef}$

E_{oed} – Oedometric modulus of deformation;
E_u – Modulus of deformation under undrained conditions;
v_u – Poisson's ratio for undrained conditions (for $Sr = 1$; $v_u = 0.5$);
E_{def} – Modulus of deformation for drained conditions;
v_{ef} – Poisson's ratio for drained conditions.

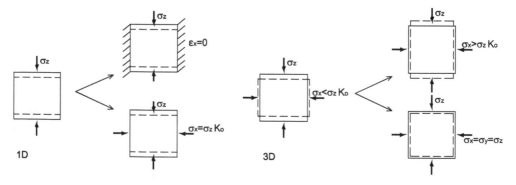

Figure 3.27 Standard types of soil element deformations

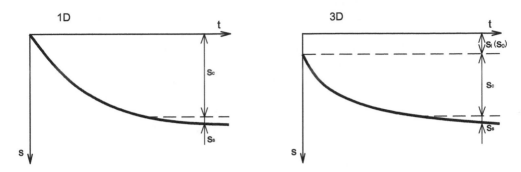

Figure 3.28 Settlement in time for 1D and 3D conditions

Settlement in time for 1D and 3D conditions are shown in Figure 3.28, from which it is obvious that for saturated soils and undrained conditions, the settlement for 1D is equal to zero – as water in pores is uncompressible. For 3D conditions, the initial settlement (s_i) is higher than zero, as the element can deform (compress) in one direction and expand in others.

Condition for fulfilment of the limit state of serviceability can be written in the form:

$$s \leq s_{\text{lim}} \tag{3.19}$$

Where the calculated settlement s is compared with an allowable limit s_{lim}, or in the form of:

$$\Delta s/L \leq \Delta s/L_{\text{lim}} \tag{3.20}$$

Where the differential settlement is compared with an allowable limit.

Owner of the transport infrastructure usually defines allowable limits. However, for runways the international standards should be fulfilled, defined by the ICAO (International Civil Aviation Organisation).

EC 7 expresses this condition in the following form:

$$E_d \leq C_d \tag{3.21}$$

Where:
E_d is the design value of the effect of actions;
C_d is the limiting design value of the effect of an action.

3.4.4.1 Short-term ground deformation

Short-term deformation is usually calculated only for fine-grained soils. Subsoil and fill can be divided into horizontal layers. In a chosen point M (usually in the middle of the individual selected layer), the initial settlement is calculated:

$$s_{i,z} = (z_i \times \Delta\sigma_{z,i})/E_u \tag{3.22}$$

Where:
z_i is the thickness of the selected layer i;
$\Delta\sigma_{z,i}$ is loading increase in the vertical direction for selected point M;
E_u is undrained modulus determined for expected range of loading for selected point M.

Total settlement is the sum of settlements of individual layers.

From the aforementioned, it is obvious that the calculation of initial settlement is relatively complicated and the predictive value is relatively low. The reason is that under the embankment axis the 1D conditions are valid, therefore the settlement for saturated soil is close to zero there. This fact can explain some results of short-term surface measurements, when the settlement for the centre of the embankment is lower than at its edges. Therefore, it is more useful to determine the settlement when the owner takes over the finished structure. As there can be a significant time delay between the end of filling and the time of structure takeover, the additional settlement can be significant as well, particularly for more permeable subsoil.

Note: The equations for the deformation of the elastic half space are sometimes used for the calculation of the spread foundation's initial settlement. However, this theory can be hardly applied to embankment settlement because of the relatively large width of the contact of this embankment with subsoil.

3.4.4.2   Long-term ground deformation

Long-term deformation of the subsoil due to construction of the embankment is much more important than short-term deformation. This value will affect the decision of whether subsoil improvement is needed. However, it is only the first step. More important is knowledge of the additional settlement of the transport line surface. Most critical is to know what the additional settlement will be after the final arrangement of the transport line surface (pavement, rail, runway). Therefore, for the second step, we need to know:

- What the degree of consolidation will be at the time of construction takeover by the investor (owner) from the contractor.
- What the impact is of secondary consolidation (value of s_s) after the end of primary (filtration) consolidation (Figure 3.28). Usually this value is very low, but for some materials, e.g. organic clays, this value should be evaluated as well:

$$s_s = C_\alpha . H . \log . t_1/t_2 \tag{3.23}$$

Where H is subsoil thickness and C_α the secondary compression index.

Long-term deformation of subsoil at the end of primary consolidation is usually calculated with the help of deformation parameters valid for 1D deformation as E_{eod}. For a relatively wide embankment, this condition can be assumed, first close to the embankment axis. Total (consolidation) settlement is after that:

$$s_c = \Sigma \Delta\sigma_{z,i} \times z_i / E_{oed} \tag{3.24}$$

For the case of symmetrical trapezoidal loading (simulating cross section of embankment), the theory of elasticity is used for load increase in a selected point M, see Figure 3.29. Kézdi (1964) provides the following equation:

$$\sigma_z = \frac{\sigma_{OL}}{\pi} \left[(\alpha_1 + \alpha_2) + \frac{a+b}{b}(\delta_1 + \delta_2) + \frac{x}{a}(\delta_2 - \delta_1) \right] \tag{3.25}$$

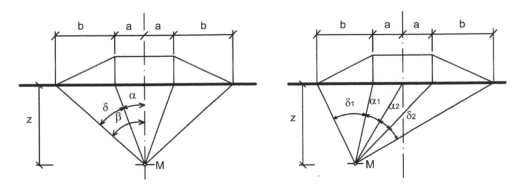

Figure 3.29 Load increase in selected points below embankment

Which is reduced for the axis of the embankment:

$$\sigma_z = \frac{2\sigma_{OL}}{\pi}\left(\beta + \frac{b}{a}\delta\right)$$ (3.26)

The graphical result of the subsoil total settlement has a typical saddle shape, Figure 3.26b.

However, Equation (3.24) overestimated the settlement in most cases. Havlíček (1978) recommended not counting a load increase lower than 10%–40%, when the oedometric modulus E_{oed} is determined in a classical way. He proposed the equation:

$$s_c = \Sigma\,(\varDelta\sigma_z - m.\sigma_{or}) \times z_i/E_{oed}$$ (3.27)

Where m is the so-called coefficient expressing structural strength ($m = 0.1$–0.4, according to the type of soil) (e.g. Vaníček, 1982; Vaníček and Vaníček, 2013b). Especially for shallow foundations, there is very good agreement between calculated and observed settlement.

This simplified approach is in general agreement with many laboratory results, where much higher stiffness was measured for small strain than for larger strain (e.g. Atkinson, 1993). The determination of higher stiffness for small strain requires very careful sample collection and the ability to test very small strains during lab tests. As strains typical for earth structures are generally higher than for other geotechnical structures – e.g. group of piles, retaining walls – a simplified approach is reasonable.

Nevertheless, better results for the deformation prognosis can be obtained when the so-called stress path control method is implemented (Lambe and Whitman, 1969). For each element in subsoil (where deformation should be calculated), stress paths are calculated for expected load change. In the lab, in a triaxial apparatus, the sample is first reconsolidated on the initial stress conditions. Consequently, the sample is loaded under calculated stress path (stress changes). Relative deformations (strains) are subsequently attributed to the layer *in situ* from which the sample was obtained.

Note 1: Vertical total settlement of embankment (fill) usually has a lower impact, particularly as the embankment is constructed from low-permeability soils (clays). Also, the height of the embankment and good compaction play important roles. However, when needed, the 1D assumption can be applied with some simplification.

Note 2: Calculation of the horizontal deformation for higher embankment is useful, namely for the comparison of the measured and calculated values to ensure that the horizontal deformation is a natural effect of the 3D deformation and not a signal of landslide development. In this case, the 3D approach is applied:

$$\Delta\,\varepsilon_x = 1/E_{def}\,(\Delta\sigma_x - v\,(\Delta\sigma_z + \Delta\sigma_y)) \tag{3.28}$$

3.4.4.3 Deformation for demanded degree of consolidation

To limit additional deformation after earth structure construction (fixing motorway pavement, rails for railways etc.), the demanded degree of consolidation can be specified for the end of construction. The curve of settlement with time should be determined in this case and:

- For 1D: $s_t = s_c \times U$ where U is the degree of consolidation (for 1D – Terzaghi's theory);
- For 3D: $s_t = s_i + s_c \times U_s$ where U_s is the degree of consolidation under triaxial conditions.

Degree of consolidation U from Terzaghi's theory is a function of T (time factor), and the relation of $U = f(T)$ is a part of the basic soil mechanics textbooks.

Time factor T:

$$T = c_v\,t/H^2 \tag{3.29}$$

Where c_v is the coefficient of consolidation (determined for the given soil) and H is consolidation length (path). Consolidation length H corresponds to the thickness of a unilaterally drained layer, respective to the half of its thickness for a bilaterally drained layer.

Very often, the speed of filing, allowing consolidation during the construction phase, is highly important. In the case, when natural consolidation is not able to fulfil the demanded value of the degree of consolidation, some measures for speeding up the consolidation process should be implemented. This is described in the next section.

3.4.5 Specificity of soft (weak) ground

Different countermeasures can be selected to guarantee the fulfilment of both limit states – ULS and SLS. Nevertheless, there are two basic opportunities:

- To improve strength properties of given soil via decreasing excess pore pressure and at the same time speed up consolidation;
- To add some reinforcing element, and at the same time to decrease deformation of the ground.

Note: Just to reduce the stress increase caused by fill from classical soil, new products can be applied as well. Lightweight aggregates from expanded clay or expanded polystyrene are such examples. However, they are used only for special cases, for example for the contact of the fill with bridge structure, either to reduce subsoil settlement or reduce earth pressure on the bridge abutment (Frydenlund and Aaboe, 1994; Vaníček and Vaníček, 2008). Waste or recycled waste materials such as ash (fly ash, pulverized ash) or foam glass are serving the same purpose as expanded clay or polystyrene. These two materials will be discussed in Chapter 4.

3.4.5.1   Pore pressure excess control – speeding up ground consolidation

The starting position is based on a calculation of pore pressure increase due to rapid loading – quick embankment construction. When this pore pressure increase is higher than allowable for ensuring slope stability, countermeasures have to be applied. The simplest possibility is just to slow down the speed of embankment filling, Figure 3.30.

Gradual surcharge is a possible solution for a relatively high embankment when the first step is applied before winter (when filling should be stopped anyway) and the second step during spring. This countermeasure is typically employed when excess above allowable pore pressure is limited and slope stability is the main problem.

In most cases, the SLS is a critical factor, and in this case additional settlement after finishing the surface of the transport infrastructure must be limited. Vertical drains are the best solution for this problem. A major aim of vertical drains is to shorten the consolidation path H, as consolidation speed is proportional to the square of this consolidation

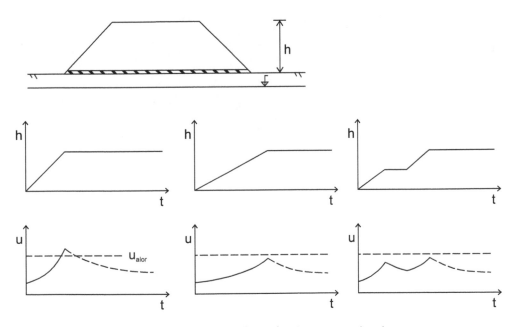

Figure 3.30 Reduction of excess pore pressure by prolonging construction time

path. A classical sandy gravel vertical drain with a diameter of about 300 mm can be used (Figure 3.31).

Nowadays geodrains are preferred, due to technical and economic advantages. Geodrains, also called wick drains or band drains, are a composite of a strip of a highly permeable core, with a width of about 100 mm and a thickness of about 5 mm, encased in filter geotextile. The geodrains are inserted into subsoil with the help of a guiding steel sleeve. The force needed for the insertion should be derived during the ground investigation phase, usually with the help of penetration tests.

The design of vertical drains consists in determining the distance between them, as this is double the length of the consolidation path in the horizontal direction. Because this distance

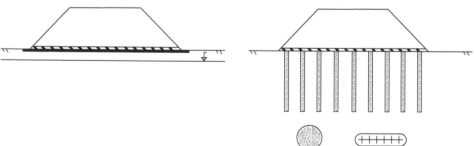

Figure 3.31 Vertical drains for speeding up consolidation

should fulfil the demands of the investor (owner), with respect to the maximal additional settlement (s_{lim}) after a certain time t after the end of filling. This time usually corresponds to the time when the surface arrangement is finished and all transport structures can be used for their expected purposes.

Demanded total degree of consolidation U for demanded time is after that:

$$U = (s_c - s_{lim})/s_c \tag{3.30}$$

When:

$$(1 - U) = (1 - U_v).(1 - U_h) \tag{3.31}$$

U_v is calculated with the help of 1D theory of consolidation, as water is drained only in a vertical direction. Consolidation path corresponds to the thickness layer for which total settlement s_c was calculated (usually corresponds also to the length of the vertical drains). After that, the value of U_h is determined.

Note: Both the degree of consolidation in vertical and horizontal directions (c_v and c_h) should be evaluated in advance.

Demanded distance (span) D between vertical drains is calculated from the Kjellman-Barron equation (see Kjellman, 1948):

$$t = \frac{D^2}{8c_h}\left[\ln\left(\frac{D}{d}\right) - \frac{3}{4}\right]\ln\frac{1}{1-U_h} \tag{3.32}$$

This equation can be solved either with the numerical calculation model or with help of a graphical solution, with the help of nomograms. These nomograms are usually part of the specifications of the individual producers of the vertical drains together with equivalent drainage diameter of specified drain.

3.4.5.2 Ground and contact reinforcement

Soil reinforcement used for soil strength improvement is a possibility when an embankment over weak subsoil has to be constructed. Reinforcing elements can be placed either on the contact between subsoil and embankment and/or within the embankment soil body.

Reinforcement of the contact is from different aspects a better solution, as it solves the stability of the whole embankment, i.e. USL, and at the same time helps to redistribute stresses on the subsoil and therefore lowers differential settlements, i.e. SLS. For contact reinforcement, a high tensile strength geosynthetic is generally applied. With respect to this reinforcement, one or two layers of reinforcement, either straight or wrapped around at the edges, can be applied. Arrangements at the edges depend on the pull-out capacity required. Most often the arrangement is as a so-called reinforcing cushion – two layers of reinforcement with wrapped around edges, Figure 3.32.

Reinforcement of the embankment body is applied when the need for reinforcement is discovered later on, e.g. based on monitoring of pore pressures, or when it is needed anyway to reinforce the embankment body itself, e.g. when slopes are designed as steep.

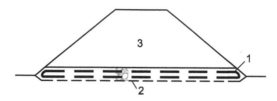

Figure 3.32 Contact reinforcement with reinforcing cushion: () contact reinforcement, (2) levelled subsoil, (3) embankment body

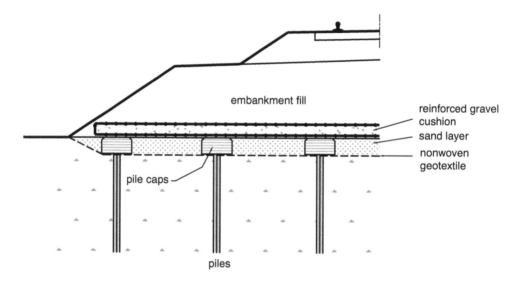

Figure 3.33 Arrangement of piled railway embankment with load transfer platform

When stability cannot be easily solved just by the contact reinforcement, or when the settlement criteria are difficult to guarantee, the solution could be an embankment on vertical bearing elements (piles, columns made of discreet subsoil improvement of different kind) Holeyman and Mitchell, 1983; Holm *et al.*, 1983). When the subsoil is very weak and *in situ* construction of any kind of column would be difficult to control, special application of geotextile-encased columns can provide the solution (Raithel *et al.*, 2004; Kahyaoğlu and Vaníček, 2019). From the embankment surface deformation (SLS) point of view, it can be determined if a load transfer platform above the vertical elements is needed. Load transfer platforms are generally made of two layers of high tensile strength geogrids in both directions (longitudinal and transversal), Figure 3.33. Several calculation models are available for the design of load transfer platforms. An overview of the most used methods in Europe was presented by Alexiew (2005) and Vaníček (2018). Those methods are also part of national codes and design guidelines – BS 8006 (UK), EBGEO (D) and CUR226 (NL).

3.4.6 Observational method design

Observational method is one of the possibilities for designing geotechnical structures. In fact, it is much more a process than a method, as the design should follow some basic principles and steps. This process is preferred when the prediction of a geotechnical structure behaviour is difficult.

The design is performed for the most probable expected input data, with the help of aforementioned analytical or numerical methods using the calculation model. Subsequently, for selected input data, acceptable limits of behaviour that can be monitored are established. However, at the same time additional measures have to be prepared for cases when the structure behaves outside of acceptable limits. Therefore, very precise plan of monitoring should be executed, one that identifies whether the actual behaviour lies within acceptable limits. If it does not, then the measures prepared in advance are applied. This process can reduce uncertainties and costs, as very often the most probable input data selected are a bit conservative.

The most sensitive input data to determine are pore pressures. Accuracy of initial excess pore pressures is very sensitive to the determination of pore pressure coefficients A and B, if Skempton and Bishop's relationship is used. Also, pore pressure dissipation with time (consolidation) is highly dependent on the coefficient of consolidation.

Precise measuring of pore pressure in subsoil under a constructed embankment is of paramount importance, as it influences the possible measures available for an observational method approach. One of the measures is to control the speed of embankment construction (filling) to guarantee that the pore pressure is lower than the critical (assumed) one. If this measure is not possible or gives unsatisfactory results, some additional measures, such as a loading berm at the toe of the embankment slope or embankment reinforcement, are added.

For optimal monitoring of slopes of cuts, instead of the pore pressure measurement it is more helpful to use an inclinometer measurement, as it is able to detect possible slip surface development.

3.4.7 Design quality control

Design quality control can reduce the risk (uncertainties) associated with geotechnical structures. There are two levels of design quality control.

The first level should guarantee that the designer is an appropriately qualified and experienced person for geotechnical structures that are classified into appropriate geotechnical category. He/she should also be able to ensure adequate communication (cooperation) among the persons involved in data collection, design, construction and monitoring. Usually some legal conditions are defined in each country. In the Czech Republic, it is building law or law concerning the chartership of civil engineers and architects. In the near future, some common designer qualification levels can be expected in Europe. New proposals are part of the draft of the second generation of the EC 7.

The second level interrelates with design control. Some consulting companies define an internal control system before the final version of the design is approved. However, for control effectiveness, the design should be controllable. For example, all presumptions and input data together with detailed specification of any software should be stated very clearly in the geotechnical design report to allow quality checking.

For important projects, some countries define additional demands on third-party checking. For example, Germany has a four-eye principle check, when a chamber of engineers appoints independent checking engineers for geotechnics. Similarly as for structure analysis, fire protection and building services (HVACR), see e.g. Katzenbach (2016).

3.5 Geotechnical design report

The final version of the Geotechnical design report serves as the basic material for the official design approval (building licence), and in some cases for the bidding process for structure realization as well. The extent – the level of details – of the GDR also depends on the structure and ground complexities from their mutual interaction – generally speaking on the geotechnical category.

The introductory section summarizes previous steps – previous design levels – with the main focus on the description of the proposed structure, including actions, description of the site and surroundings, up-to-date experiences with similar structures and ground conditions, evaluation of the suitability of the site with respect to the proposed structures and potential risk.

Subsequently evaluates the results of the geotechnical/ground investigation report that can be in GDR annexes. The results of GIR and GDR annexes mainly provide a detailed description of the ground conditions, expected interactions between the proposed structure and the ground, and finally an evaluation of the results of the field and lab tests, which include a summary of measured or derived geotechnical parameters values.

The geotechnical design model indicates characteristic values of the geotechnical parameters, including the approach selected for their determination together and a justification of their final selection with respect to the solved limit states.

The calculation model contains the most important part, verifying structure safety and serviceability, with respect to the potential limit states, design situations and relevant actions. The geotechnical design calculations are supplemented with statements on the codes used, standards and literature, and also with drawings. An integral part of the design is the technology of the geotechnical structure construction.

The geotechnical design report should also summarize proposals for structure supervision, inspection, monitoring and maintenance, together with recommended requirements on them.

3.6 Structure construction

Personnel having the appropriate skill and experience should carry out structure execution according to the relevant standards and specifications.

For earth structures of transport engineering, the most relevant construction standards were prepared by CEN/TC 396 Earthworks, namely:

- EN 16907–1, Earthworks – Part 1: Principles and general rules;
- EN 16907–2, Earthworks – Part 2: Classification of materials;
- EN 16907–3, Earthworks – Part 3: Construction procedures;
- EN 16907–4, Earthworks – Part 4: Soil treatment with lime and/or hydraulic binders;
- EN 16907–5, Earthworks – Part 5: Quality control.

Parts 1 and 2 should help to guarantee that the engineered fill is suitable for a given part of the earth structure. Part 3 is focused on acceptable placement, water content, compaction requirements and fulfilling geometric demands. Part 4 deals with cases of soil improvement, namely with soil stabilization with lime and/or other hydraulic binders. They are considered a group of standards for executing earthworks. Earthworks are a civil engineering process aimed at creating earth structures.

The final phase of earth structure construction has, therefore two basic levels:

• To approve that all assumptions and demands specified during the earth structure design were fulfilled – control of design assumptions;
• To approve that all demands specified in standards for executing earthworks were fulfilled – workmanship control.

Plans for structure supervision (including quality control), inspection, monitoring and maintenance create the framework for such control. Practical realization requires cooperation of basic partners – representatives of the contractor, designer and investor/owner/client.

3.6.1 Checking of design assumptions

The basic check of design assumptions consists of verifying the data used for the structure design:

• Verification of ground model and the respective geotechnical design model;
• Verification of the fill quality, whether it corresponds to the geotechnical parameters used in the calculation model.

3.6.1.1 Geotechnical design model check

As already mentioned, the check for geotechnical structures, namely for earth structures, is much more problematic than for other structures. For subsoil of embankment, the checking is practically zero, as only the top part of the subsoil – after removing organic ground – is visible. For cuts, there is a better chance, as outcrops enable the comparison of reality with assumptions. The case of proven difference, e.g. in the form of significant discontinuity such as the presence of a small layer with different properties – needs a complementary ground investigation. After the geotechnical design model correction, a new verification of the limit states should follow.

3.6.1.2 Compaction control

As also already mentioned, the control of the geotechnical parameters used in the design is still performed indirectly via compaction control. Designer, in agreement with Figure 2.4, recommends characteristic values for the fill on the basis of the lab compaction tests. Therefore, he/she specified the range of moisture content and dry density for which lab tests of mechanical-physical properties were performed and finally characteristic values were recommended. The main aim of compaction control therefore consists in the control of these values – moisture content w and dry density ρ_d. For coarse soils, the impact of moisture content is very small, and the control of dry density is therefore sufficient. Obtained dry density is compared with maximal dry density $\rho_{d,max}$ from the Proctor compaction curve (for fine soils, generally for soils sensitive to moisture content). The ratio of these two values is denoted as D.

$$\rho_d/\rho_{d,max} = D \qquad (3.33)$$

Where D is degree of compaction and often expressed in percentage. Most often, the demanded degree of compaction is between 90% and 100%. The term compaction to 95% PS (Proctor standard) means a required ratio of $\rho_d/\rho_{d,max} \geq 0.95$.

Note: Proctor tests in the lab are performed on particles lower than 5 mm respective 16 mm. Therefore, the recalculation of dry density and optimal moisture content is needed for specified percentage of coarser grains.

For coarse soils, obtained dry density is recalculated on voids ratio e. Subsequently, the obtained value is compared with maximal and minimal values obtained by a prescribed lab test. Requirement of their compaction rate is based on the relative density index I_d:

$$I_d = (e_{max} - e)/(e_{max} - e_{min}) \qquad (3.34)$$

Demanded value of the relative density index I_d is around 0.9 and higher.

For soils sensitive to moisture content, the required range of moisture content Δw is specified as:

$$w = w_{opt} \pm \Delta_w \qquad (3.35)$$

Where Δw is a value around 1.5%.

The previously described compaction control via dry density, respective moisture content, is still the method used most often for standard earth structures. Nevertheless, there are some questions about the effectiveness of such a procedure:

* How quickly the results of performed control can be obtained;
* How many control samples are needed, when a frequency of control is very often prescribed by national standards or by main investors (owners), such as national motorway or railway authorities;
* How many sample results can be accepted when they are not fulfilling the demanded values, and what difference from demanded value is still accepted;
* How poor compaction can be corrected.

Even when the aforementioned questions practically remain, the preference, especially for earth structures of transport infrastructure, is now given to compaction control via determination of the deformation modulus E_{def}. With the help of E_{def}, the conditions for motorways and highways classified into different categories can be prescribed. This applies similarly for railways lines when the higher speed of trains requires a higher deformation modulus.

The static load test using a small loading plate (diameter ca 0.3–0.37 m) is presently the most common type of testing of the deformation characteristics of compacted layers, Figure 3.34.

Successively, gradually, the surcharge at the contact of the loading plate with the subsoil is increased and vertical deformation y measured after setting. Two loading cycles with maximal value of about 0.2 MPa and with unloading up to zero load are a typical case. The

Figure 3.34 Static load test arrangement

modulus of deformation for the second loading cycle is the main output. When using a simplified relation:

$$E_{def,2} = \frac{1.5 \times r \times \Delta\sigma}{y_{max,2} - y_{0,1}}$$

(3.36)

Where r is plate radius and $\Delta\sigma$ is the range of plate loading for which the range of plate settlement was measured.

A supplemental condition is expressed as the demanded ratio of two measured moduli (for the first and second loading cycles), with a recommended value of about 2.2 for fine soils and 4.0 for coarse soils. Nevertheless, the static plate load test is becoming a kind of index test – it is fast to carry out, easy to define and has considerably high credibility, allowing comparison of individual results.

In order to apply the deformation moduli obtained from the static plate load test for the subsequent calculation of the earth body deformations, the geotechnical engineer must show a great deal of foresight (Vaníček and Vaníček, 2008).

The same applies, even to a greater extent, to "dynamic plate load tests" (Figure 3.35).

A very important step forward are vibratory rollers, which are equipped with so-called continuous compaction control (CCC). A big advantage of this method is the direct control

Figure 3.35 Dynamic load plate test arrangement

during, but not after, the compaction process. Brandl *et al.* (2005) presented a highly detailed description of the CCC system, Figure 3.36.

The vibration detector (A) measures the response of the compacted layer to the passes of the vibratory roller. A processor unit (B) converts the signals received into measured data. Subsequently the display and memory unit (C) continuously displays and documents the measured data. Finally, a sensor (D) records the roller's velocity and the distance travelled. When the results of the measured response are connected with place, the driver can see the results for that place. If the measured data of the vibratory response are verified in advance

1)

2)

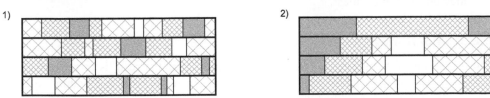

Figure 3.36 Components of CCC systems: (a) transducer A, processor B, memory C, location sensor D – GPS based speed; (b) driver's display – control plots of two compacted zones

on the testing locality for the same material and manner of placement, the driver can easily recognize where the ground is well compacted and where it is not. The result is uniform ground compaction, strictly limiting differential settlement, which is even more important than the value of total settlement.

With CCC systems, consumption of energy can be significantly reduced. Some detailed specifications of this method can be found in Chapter 4.

3.6.2 Workmanship control

A good standard of workmanship has a great positive impact on the final structure behaviour. The first step relates to the earthwork design – with proposals of construction technology. Significant guidance in this direction are standards prepared by CEN/TC 396 Earthworks, namely "EN 16907-3, Earthworks – Part 3: Construction procedures." When realizing an earth structure, final control performed by all partners guarantees that the structure was constructed in agreement with initial demands.

The intention of this section is not to go into great detail but to summarize basic principles. *Before the construction starts*, the main focus should be on:

- Excavation of material from the borrow pit – generally from the place from which soil will be used – followed by the manner of transport, all with respect to climatic conditions expected on the site to guarantee that the moisture content will fall within the demanded range.
- Recommendation of compaction equipment types (different rollers are the most important), and specification of thickness of compacted layer and number of passes – on the basis of previous experience (for simple structures with limited volume of deposited material) or on the basis of trial tests for more significant structures.

The same is valid either for nonstandard materials or for fill improvement – e.g. specification of technology of soil mixing with a binder and the manner of spreading and compaction. *During the earth structure performance*, the main focus should be on:

- Adjustment of the level on which the excavation will end or filling will start and its protection against climatic impacts.
- Control of the recommended material spreading and compaction technology, including compaction of embankment edges.
- Control of the compaction results.
- Embankment reinforced by geosynthetics requires specific conditions – see also Chapter 4 concerning soil nailing for slopes of cuts improvement and generally for any other methods of improvement.

3.6.3 Monitoring

The main aim of monitoring, performed in agreement with the plan, is to prove that the designer's assumptions under which the limit states were assessed fall within expected ranges. Monitoring is usually focused on deformation and pore pressure measurement. Deformation measurement takes different forms, e.g. vertical or horizontal measurement of the earth structure contours or vertical inclinometers for monitoring potential development of the slip plane and horizontal inclinometers for settlement monitoring of the cross section contact between ground and fill.

Monitoring has in principle two phases:

- At the end of construction, the monitoring results should prove that the structure is finished under the demanded conditions and that the owner can take over this structure and start using it.
- During the guarantee period, the monitoring results should prove that the time depended influential (pore pressure, deformation) are in accordance with the design.

For very important earth structures only, monitoring can continue up to the end of the structure's life expectancy.

Visual monitoring is typical for simple earth structures as well as for surface erosion monitoring.

3.7   Maintenance

The advantage of earth structures over structures from man-made materials – concrete, steel, timber – is higher resistivity to deterioration, especially when they are constructed with the demanded safety. The problem of structure maintenance is now in the centre of interest, namely when specifying the bidding price. This price can be defined at the end of construction (at the time of structure takeover by the owner) or for the service life of the structure. For example, the second price allows the new possibility of comparing a high embankment with a bridge structure. Nevertheless, in all cases, a good design from the beginning can significantly decrease the cost of maintenance.

For transport engineering earth structures, there are two typical problems.

- Maintenance of the structure surface, consisting of securing its vertical position. Problems are associated with repeated variation in stress (dynamic loading by transport), namely with a high frequency or a high ratio of this stress increase to geostatic stresses. As this is connected with pavement for road structures or with superstructure for railways, this problem is in the hands of road or railways engineers. Typical problems in this area are connected with the corrugation of pavement close to the bus stops or with longitudinal footprints caused by heavy lorries.
- Maintenance of earth structure slopes – e.g. planting of greenery, repair of surface erosion.

The surface issue can be attributed to the poor quality of either the pavement/superstructure or the subbase. The main cause can be usually visually determined by the shape of the surface irregularities, therefore the repair can be designed accordingly.

3.8   Geotechnical construction record

The Geotechnical construction record (GCR) compiles all documentation of construction, supervision, inspection and monitoring for each execution phase as well as for the final state. GCRs create a certain experience for similar future structures. The last version of the geotechnical/ground model, supplemented by the contours of the realized structure, is very useful for the next activities, as:

- Can assist with future maintenance, design of additional works and for the phase of structure decommissioning.
- Source of information for future activities – e.g. with transport infrastructure widening, or for evaluating the interaction between the existing structure and a new structure.
- Source of information when during natural or man-made hazards require quick action (flood, accident of truck with chemical products).

As all of this information is compatible with information concerning building information modelling (BIM), the last section of this chapter is devoted to that subject.

3.9   BIM and geotechnical engineering

The term BIM is relatively new, as can be deduced from its simplest definition: "BIM is a common name for the digital representation of the building process." According to Wikipedia, the term "building information model" was used for the first time by van Nederveen and Tolman (1992). In the UK, this term is defined as a process for creating and managing information on a construction project across the project lifecycle. In the Czech Republic, Vaněk (2011) describes BIM as an information database which can encompass all data starting from the design through the construction phase, management facility and maintenance to the structure's demolition. Therefore, it encompasses all the information utilizable through the project's life span. It is necessary to add that all partners of the construction process contribute to this information database. BIM cannot be interpreted, however, as only an information model or information database. It must instead be seen as an important approach to construction design and management.

From these short definitions, it is clear that definitions of BIM demands are emphasizing:

- Good cooperation of all partners of the construction process;
- A sustainability approach, taking into account all life expectancy;
- Top-quality software enabling the creation of the complex information model.

From these basic points, it is clear that the decision to apply BIM should start at the beginning of any construction process. Geotechnical engineering is responsible for ground and foundation structures, on which all other structures are constructed. Therefore, any BIM process should start with information regarding ground and foundation structures.

For other geotechnical structures, such as underground and earth structures, the situation is little bit different. These structures are also based on ground but are more independent. Underground transport infrastructures, such as transportation tunnels or metro, are individual constructions for which geotechnical engineers are responsible, particularly from the safety and serviceability point of view. Even when the deterioration of the tunnel lining (lining ageing) is significantly slower than that of the inner equipment, especially the security system, knowledge of the lining's deterioration curve is very useful for defining the most appropriate time for maintenance or repair (Soga *et al.*, 2011). Earth structures for transport engineering are in very similar situation. Good cooperation with transport engineers is a basic assumption. Deterioration due to ageing of pavement or the superstructure, as well as the security and information systems, is much quicker than the deterioration of the earth structure. Therefore, underground and earth structures are central to the BIM process.

3.9.1 BIM principle

BIM starts from the assumption that up-to-date praxis, represented by different sets of 2D and 3D drawings supported by different technical reports and calculations, insufficiently covers global problems of the construction process. BIM consists of the realization and subsequent safekeeping and utilization of the 3D model of construction. This model is created by individual parametric elements holding specific information, which can be utilized within the lifecycle of the project – starting with the phase of preparation (investigation and design) and continuing with realization and maintenance and ending with construction disposal.

From the aforementioned, the following statement can be made:

- Geotechnical structures as an independent tunnel or selected section of motorway or railway can be a starting point of the individual BIM model with individual parametric elements around, Figure 3.37a.
- Geotechnical structures can create an individual parametric element which is connected to the basic BIM model, Figure 3.37b. This is typical for foundation structures, where building creates the basic BIM model. But it is also the case of tunnel or earth structure for transport engineering, when they are an independent element of the basic BIM model of the proposed motorway, Figure 3.38.

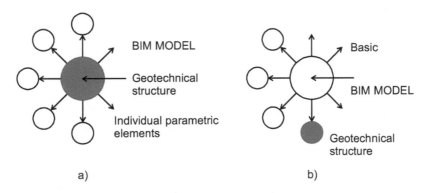

Figure 3.37 The position of a geotechnical BIM model for different geotechnical structures: (a) geotechnical structure creates a basic BIM model; (b) GS creates an individual parametric element around basic BIM model

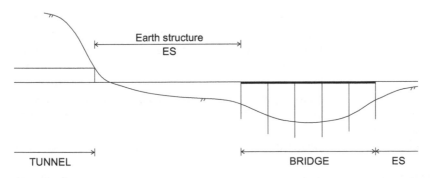

Figure 3.38 Tunnel, bridge and earth structures are elements of the BIM model of the proposed motorway

3.9.2 Application of BIM for geotechnical structures

Figure 3.39 depicts a 3D basic BIM model of the geotechnical structure – embankment of transport infrastructure. This basic model has several individual parametric elements directly connected with this geotechnical structure as well as several non-geotechnical ones, such as pavement, security and transportation information systems.

The advantage of these geotechnical elements is that they are in agreement with individual models demanded by the Eurocode 7 geotechnical design:

- Geological and geotechnical models development, where individual steps of the geotechnical/ground investigation report (see Table 3.1) relate to the individual steps of the design.
- Database of obtained measured and derived geotechnical parameters for individual lithologic ground layers.
- Geotechnical design model.
- Calculation model.
- Phase of structure realization – with specification of geotechnology and with results of construction (e.g. compaction) control.
- Detail specification of used structural elements – e.g. vertical geodrains, reinforcing geosynthetics etc.

But we are still speaking about a basic 3D model. Often, some additional dimensions are discussed.

- Fourth dimension: time – playing a very important role for geotechnical structures.
- Fifth dimension: financial aspects – e.g. demanded cash flow.
- Sixth dimension: bidding process – helping to select the most effective proposal, including long-term maintenance and respecting a sustainability approach.
- Seventh dimension: timetable of structure maintenance and improvement, including the affordability and availability approaches.

From the early development of the BIM process, different applications were offered from the main software suppliers. These programs enable parametric modelling, and generally they support BIM. But up to now there has been no software able to guarantee all BIM functions. Nevertheless, BIM functions are dealing with a set of applications and instruments which communicate with each other, guarantee different inputs, however they are arising from common source database. So this means that one platform supports communication between individual applications. In this respect, the platform IFC (Industry Foundation Classes) is mentioned; in the Czech Republic the most used platform is Revit from Autodesk.

One such BIM application used in geotechnical engineering is HoleBASE SI, which represents a complex geotechnical system for data management. The application enables the user to create protocols, graphs, reports and interpretations in just a few seconds. Within the framework of this application, it is possible to create basic ground and geotechnical models. A similar system is gINT.

It is obvious that the transfer to the BIM model will be continuous, whereas any experiences from realized applications can be very useful for the future. In this respect, two examples can be mentioned: the Silverton Tunnel in London (Moryn, 2016) and the Bergen

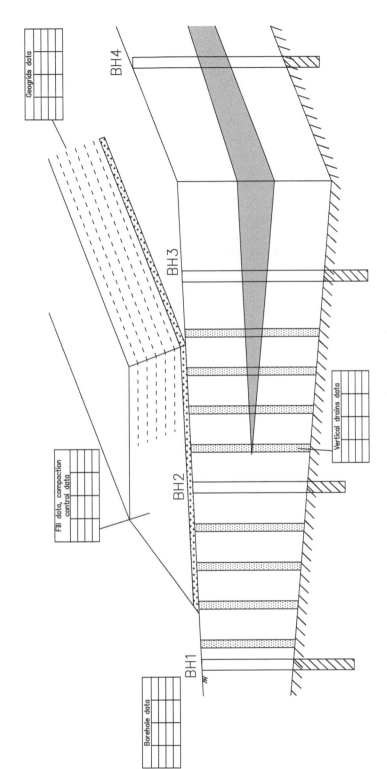

Figure 3.39 Basic 3D BIM model of the geotechnical structure — embankment of transport infrastructure

Light Rail in Norway (McInerney, 2016). Very useful information was obtained from a very complicated structure – a new sewage treatment plant in Prague, which is situated below ground level (Kuba and Wallensfeld, 2019).

To conclude, some advantages of the BIM 3D model can be mentioned for future utilization. Particularly for earth structures for transport infrastructure, the model can be used when:

*   Structure widening is proposed, e.g. from four lines to six lines;
*   A new transport infrastructure (including different pipes) will cross the existing one;
*   A natural hazard will hit an existing structure;
*   Human-caused accidents will affect an existing structure – e.g. accident of a truck with chemical product will need quick action to decide what remediation method should be used where the accident occurred.

Therefore, model safekeeping and access to it for quick utilization are preconditions of high functionality.

Chapter 4

Sustainability design approach

The concept of sustainable development was accepted at the Rio de Janeiro Earth Summit of the in 1992. After that, this concept was gradually developed in various areas of human activities, as well as for construction sector (Mulligan, 2019; Vaníček, 2011), including geotechnical engineering (e.g. Vaníček and Vaníček, 2013a; O'Riordan, 2012; Vaníček et al., 2013; Correira et al., 2016).

As was already mentioned in Introduction, the main aim is to provide economically competitive construction with higher utility value. This higher utility value results from lower energy demands, lower raw material inputs and lower need of new plots of land when the risk of the danger for human health and life during natural disasters, accidents and unwanted events is reduced.

The sustainability approach is now included in all documents prepared for transport infrastructure, together with affordability and availability approaches.

At this point, we can state that geotechnical engineering has always had very close contact with the environment. The last few decades, however, have brought new possibilities for how to react positively to new demands. The following are some examples:

- New technologies for soil stabilization, to enable the use of less appropriate soils for earth structure construction;
- New technologies for soil reinforcement, either for fill with geosynthetics reinforcement or for cuts by applying soil nailing;
- New possibilities for compaction control via continuous compaction control methods;
- New technologies in the field of geothermal energy utilization, namely energetic foundations;
- New possibilities for structure protection against natural disasters, such as floods, landslides, rock falls;
- New knowledge in the field of contaminant movement in the ground, together with new methods of subsoil decontamination;
- New possibilities for comparing different geotechnical structures from the view of energy demands (or CO_2 footprint).

However, the acceptance of these new demands is not only in the hands of geotechnical engineers (designers, contractors) but also very strongly in the hands of investors (owners/ clients). The investor should state from the beginning that he/she also prioritizes these principles. The BIM approach can be in this respect the right option. The investor can use these new principles within the bidding process, the result of which could not only be a better technical solution from the long-term perspective, but could also lower the costs from the point of view of structure life expectancy.

The sustainability aspect accentuates increasing the resource efficiency of infrastructure transport, namely via the development of more economically and environmentally acceptable earth structures. In this chapter, we will mostly focus on saving land, natural aggregates and energy. Sometimes these approaches to saving are interconnected, as with new proposals that not only reduce the consumption of land, but also lower the use of aggregates and energy. Some discussion will be devoted to the maintenance of the earth structures, since it contributes significantly to the overall cost during the structure's expected lifetime.

4.1 Importance of accepting the sustainability principle from the very beginning

The sustainability design approach has a great chance to succeed when the bidding price will not be connected solely with construction costs. There are two very important aspects:

*   A lower construction price can lead to higher costs for maintenance, and more frequent maintenance is not in agreement with two other aspects – availability and affordability;
*   The bidding price can include savings (expressed in money) via the saving of land (as a whole or just via greenfields) and/or natural aggregates (via application of the large volume of waste which can be used as an engineered fill and not stored in landfills).

Buyout of land for new transport infrastructure is a sensitive problem, which often has a negative impact by prolonging the length of the construction preparation phase. Reducing the costs for land buyout is also very important. The question is, when should the price of saved land be counted, at the moment of buyout or, for example, for the expected price at half of the structure's life expectancy? The investor (owner) should specify this from the earliest phase of the whole process. Steeper slopes of fill or cuts are possible with the application of soil reinforcement, saving not only land but also energy for aggregate transport and compaction.

This is the same for the application of a large volume of waste as engineered fill. The vicinity of the source of this waste can change the classical requirement about the balance between volume of cuts and embankments with preference of fill.

All these new principles should be considered during the selection of a new road or railway line. Comparing alternatives of road and railway alignments is useful not only for the design itself but also for the environment impact assessment (EIA), a process which is now mandatory for significant infrastructure projects.

This comparison is useful in the earliest phase of the construction process. The comparative study can start at the end of the first phase of the ground investigation (desk study), when all information from previous projects and geo-environmental maps are evaluated. In the Czech Republic, a large set of geo-environmental maps in the scale 1:50,000 are used; for major towns, an additional set of four maps in the scale 1: 5000 provide geological, hydrogeological and superficial deposit information and documentation points.

On this level, the range of information about the ground enables the comparison of individual alternatives of transport infrastructure alignments from both the environmental point of view (during the EIA) and the engineering point of view – with the help of pre-assessed geotechnical properties of ground influencing the design of the earth structures of transport engineering.

Nevertheless, the predominant roles in this phase of project preparation are played by the investor (owner) and the designer/transportation engineer, who are responsible for whole transportation project. First, the investor should clearly specify his/her approach to the principles of sustainable construction and how the proposed solution, which fulfils both engineering and sustainability aspects, will be evaluated during the bidding process on both design and construction levels.

The geotechnical engineer has only a consultative role in this phase. Therefore, the main aim of the next chapters has two levels. The first is to a basic orientation to the sustainability approach. The second goal is no less important, as it provides technical support for how to utilize all the new possibilities that the progress in geotechnical engineering enables in this field.

The following sections covering different savings – land, natural aggregates and energy – is only estimative, as there are significant mutual cross connections.

4.2   Land savings

Consumption of land denoted as greenfields is relatively huge – for the Czech Republic, it is up to 20 ha per day. For countries like Germany and the UK, it is at least five times more. Therefore, each country is trying to define a strategy to decrease existing consumption. Usually, it is in the form of a national brownfields regeneration strategy. A basic possibility is to orientate new construction on brownfields, on ground that was previously used and now devalued. The properties of such ground were in most cases changed, mostly negatively, with lower mechanical-physical properties (e.g. more compressible) or chemical changes, with a certain degree of contamination. It is interesting to mention that there are also so-called railway brownfields. Such land is sometimes used for a new reconstructed railway system, but predominantly for other new construction.

Land savings in the field of transport infrastructure have roughly two possibilities:

- Savings during definition of the alignment of the transport infrastructure, either by situating the new alignment on brownfields or with a new profile which does not require high embankments or deep cuts (as they need more land).
- Savings due to slope inclination increase via soil reinforcement, either via geosynthetics soil reinforcement for fill or soil nails application for cuts.

As the first possibility was briefly mentioned already and is more or less in the hands of transport engineers and investors, further attention will be paid to the second possibility.

4.2.1   Geosynthetics soil reinforcement

Geosynthetics soil reinforcement is now a typical example of fill reinforcement, allowing steeper slope construction, Figure 4.1. The basic principle of soil reinforcement consists of adding to the soil tensile strength, which is generally very low for soils. Geosynthetics with a high tensile strength are able to fulfil this reinforcing function.

Different geogrids or woven geotextiles are typical examples for such an application. Different geosynthetics create a 2D reinforcing element, which is located between two individual compacted soil layers. In this case, we are speaking about macro-reinforcement. The potential slip surface either cuts the reinforcing geosynthetics or passes behind the reinforced zone, where it is kinematically more demanding, with higher degree of safety, Figure 4.2.

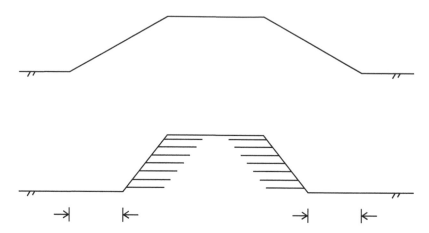

Figure 4.1 Land saving via fill reinforcement

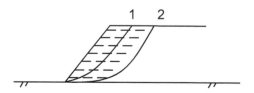

Figure 4.2 Reinforced embankment on stiff ground (1) slip surface cutting reinforcing geosynthetics, (2) slip surface behind zone of reinforcement

Another possibility of reinforcement is micro-reinforcement, but up to now it has limited practical application. French-patented Texsol (Leflaive, 1985) utilizes continuous yarns during embankment filling. Short cut-outs from 2D geosynthetics (McGown *et al.*, 1985) or short fibres about 40–50 mm in length are further possibilities of such micro-reinforcement (e.g. Michalowski and Čermák, 2003; Consoli *et al.*, 2005; Škara, 2017).

For comparison of the soil reinforcement development, we can find analogy with concrete:

*   Mass concrete – compacted soil;
*   Reinforced concrete – soil macro-reinforcement;
*   Fibre-reinforced concrete – soil micro-reinforcement;
*   Pre-stressed concrete – preloaded and pre-stressed reinforced soil, e.g. (Tatsuoka *et al.*, 1996).

Concerning macro-reinforcement, the following geosynthetic reinforcing elements properties are the most important:

*   Tensile strength (together with tensile elongation);
*   Friction along the soil and geosynthetics contact (tg φ_{sg}).

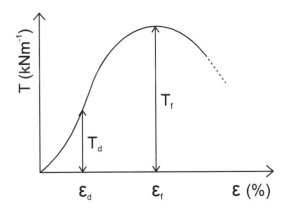

Figure 4.3 Typical loading curve for quick tensile test

T_f – tensile strength at failure, corresponding elongation ε_f;
T_d – design tensile strength, corresponding elongation ε_d.

Note: European codes for geosynthetics testing are prepared by *CEN/TC 189 Geosynthetics*.

However, a few remarks concerning short- and long-term tensile tests are useful. For a quick tensile loading test performed on a strip from geosynthetics, a typical loading curve is obtained (Figure 4.3), where T_f represents tensile strength at failure and ε_f represents relative elongation (strain) at failure (for ultimate strength). As geosynthetics are produced from synthetic thermoplastic materials, such as polyester (PES), polypropylene (PP), polyethylene (PE), polyamide (PA), polyvinyl alcohol (PVA) and aramid (AR), they are all sensitive to creep. According to Jones (1996), most sensitive to creep are polypropylene tapes and grids, followed by polyethylene of high density (HDPE) and polyester fibres, and least sensitive are polyaramid fibres. Therefore, the design value of tensile strength T_d, and respectively the design value of elongation ε_d, are also presented in Figure 4.3 – as these values are used during the design. Design strength value is about 20%–40% of the maximum peak value, depending not only on creep properties, but also on the sensitivity of the earth structure to relative deformation.

4.2.1.1   Geosynthetics soil reinforcement design

First, fill reinforcement has a positive impact on ULS GEO type. The impact of reinforcement on the SLS limit state is very small for fill; nevertheless, it has a certain positive role in the case of contact reinforcement of the fill with subsoil, namely for differential settlement levelling in cross section.

For an analytical calculation model, the principle of soil reinforcement on slope stability for a simple case with one reinforcing element is obvious from Figure 4.4 (Vaníček and Škopek, 1989; Vaníček and Vaníček, 2001).

For case (a), a reinforcing tensile force adds moment to cantilever y. Increasing cantilever y has a positive effect. For case (b), an additional moment is acting on cantilever R, so it means that the positive effect is independent of the location of the reinforcing element. Case (c) assumes that the reinforcing element is acting as an additional horizontal force, and

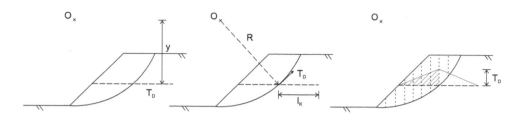

Figure 4.4 Three basic opportunities for incorporating the reinforcing element into slope stability analysis

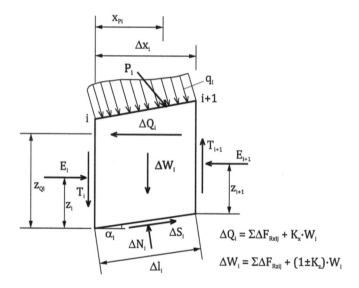

Figure 4.5 Forces acting on an individual slice

therefore its maximum is in the point of intersection with the slip surface. The highest effect for such an assumption is where the tangent line parallel to the slope surface touches the slip surface. For this assumption, the area of a triangle, which is bordered by design tensile force T_d and by length of reinforcement, is highest. It is roughly in the lower third of the slope. This assumption is the most logical and applies well to Janbu's method of slope stability as additional inter-slice horizontal force (Janbu, 1973).

Janbu proceeds from a moment equilibrium condition as well as from the conditions of equilibrium in vertical and horizontal directions. As a supplemental condition, he uses the knowledge of the position of inter-slice and normal forces on the base of an individual slice, Figure 4.5.

The conventional degree of slope stability according to Janbu is:

$$F = \frac{\sum\limits_{a}^{b} A}{E_a - E_b + \sum\limits_{a}^{b} B} \tag{4.1}$$

Where E_a and E_b are horizontal forces acting above and below a landslide segment. However, very often, as for the situation in Figure 4.4, they are equal to zero. For supplemental expressions, apply:

$$A = \tau_f \Delta x \left(1 + tg^2 \, \alpha\right) \tag{4.2}$$

$$B = \Delta Q + (p + t) tg \, \alpha \, \Delta x \tag{4.3}$$

$$\tau_f = \frac{c' + (p + t - u) tg\varphi'}{1 + \dfrac{1}{F} tg\varphi \, tg\alpha} \tag{4.4}$$

Where:
ΔQ is the horizontal force increment (mainly from reinforcing geosynthetics):

$$p_i = \frac{\Delta W_i}{\Delta x_i} \text{ and } t_i = \frac{\Delta T_i}{\Delta x_i}$$

The final result is obtained by step-by-step iteration. The authors of this volume use using their own software, SVARG (Vaníček and Vaníček, 2000).

When the solution is performed according to the demands of EC 7, design values of geotechnical parameters are used – φ'_d and c'_d.

The next step is connected to a control of the length of anchoring L_k (length of the reinforcement behind the slip surface) to guarantee safety against reinforcement pull-out:

$$L_k = \frac{\gamma_L \cdot T_d}{2 \cdot \left(\gamma \cdot h \cdot tg\varphi_{gsd} + a_d\right)} \tag{4.5}$$

Where T_d is the long-term design tensile strength of geosynthetics.

ISO/TR 20432 indicates the following equation for T_d:

$$T_d = \frac{T_k}{RF_{CR} \cdot RF_{ID} \cdot RF_W \cdot RF_{CH} \cdot f_s} \tag{4.6}$$

Where:
T_k – characteristic value of the long-term tensile strength;
RF_{CR} – reduction factor for creep;
RF_{ID} – reduction factor for damage during installation;
RF_W – reduction factor for atmospheric exposition;
RF_{CH} – reduction factor for chemical environment;
f_s – reduction factor for data extrapolation.

The numerical calculation model based on FEM is another possibility. The development in this respect is slightly complicated, as two very different components are in contact, and the reinforcing element is very thin compared to the compacted soil layer. Rowe and Ho (1988) present an overview of the different approaches.

Currently, the discrete model of reinforced soil prevails. Special elements are used for reinforcement, completed with contact elements modelling the interaction of both reinforcement surfaces with surrounding soil.

Yashima (1997) presented certain views of the FEM development for reinforced soils. The authors of this volume (Vaníček and Vaníček, 2008) summarized the development of FEM as follows:

- More general constitutive laws better expressing real soil behaviour for the modelling of stress and deformation of soil;
- Special finite elements for the modelling of reinforcing grids from geosynthetics;
- Special elements for modelling the mutual interaction of reinforcement with soil.

As for the reinforced earth structure, the state of planar deformation prevails; after that, modelling in 2D is possible. Planar elements can be used instead of 3D elements. After that, the numerical modelling is substantially simplified. Karpurapu and Bathurst (1994) compared results of the numerical model with empirical ones in the scale 1:1 and obtained very good agreement. However, they concluded that a numerical model requires not only a careful choice of boundary conditions but also a step-by-step construction process.

4.2.1.2   Geosynthetics soil reinforcement application

The application is presented on an example of a high wall from reinforced soil, which after construction was completed started to show marks of instability. On its top, a local road was proposed. The authors were responsible for remediating the structure (Vaníček and Vaníček, 2016).

Construction of the logistic and distribution centres requires a large area at the same horizontal level. This condition usually needs to situate the first part of the centre on the fill, material of which is obtained on the opposite side (cutting), where the footing bottom is situated some metres below the original surface.

In the case described, the new warehouse encompasses the total area of the construction site – 800 × 550 m. The first two halls, A and B, were finished in 2006, the third in 2010. The last one, hall D, situated on the fill, was in the preparation phase when the authors were invited by the investor to evaluate non-typical observed behaviour of the geosynthetic-reinforced retaining wall circumscribing the fill. This retaining wall along the perimeter of the fill was about 10 m in height. The finished wall will be loaded by heavy traffic on a roadway situated just behind the crest.

A large main crack as well as wall leaning were registered during the site visit after a short period of construction interruption, Figure 4.6. The main crack had a width of about

Figure 4.6 View of constructed reinforced wall and main crack parallel with upper edge of wall

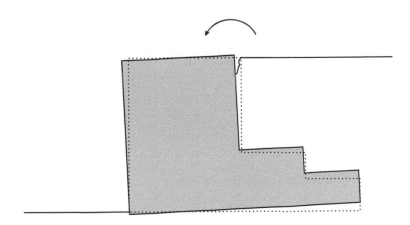

Figure 4.7 Tilting of the reinforced wall/quasi-homogeneous block

0.4 m and a depth almost 5 m and was parallel with the upper edge of the wall, with a distance of about 7–9 m. The height of the wall was about 11 m at these places. The ends of the reinforcing geogrids were visible in the main crack – on the side closer to the embankment crest.

After evaluating the existing geotechnical investigation, original wall design and its monitoring, it was concluded that the main problem was associated with tilting of the reinforced block and attributed to the limit state EQU, Figure 4.7.

Tilting was caused by differential settlement of the reinforced block. The tensile crack created behind the zone of reinforcement gradually deteriorated stability, particularly when the crack was filled by surface water.

Numerical modelling approved the first evaluation and helped to specify the main reasons for the differential settlement:

• Natural reason – 2D (3D) deformation below the front edge of the block is higher than the 1D deformation below the inner edge;
• Poorer ground conditions were observed and subsequently proved below the front edge (old brook sediments, pipe outlet for surface run-off water);
• Shear strength was exceeded in the front edge, causing horizontal displacement due to plasticization, Figure 4.8;
• Reinforced block was relatively stiff, as soil improvement by lime stabilization was applied on fill.

For the wall remediation, the following countermeasures (steps) were recommended:

• To remove fill at least to the depth of crack – roughly 5 m, Figure 4.9;
• To install vertical elements (H piles) at the wall toe to limit horizontal movement – subsoil extrusion, Figure 4.10;
• To use much longer geogrids in the upper part to increase the moment, protecting overturning.

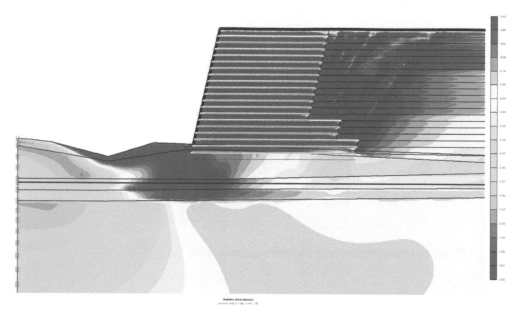

Figure 4.8 Horizontal extrusion below front edge due to plasticization

Figure 4.9 Removal of old fill with new reinforcement of upper part

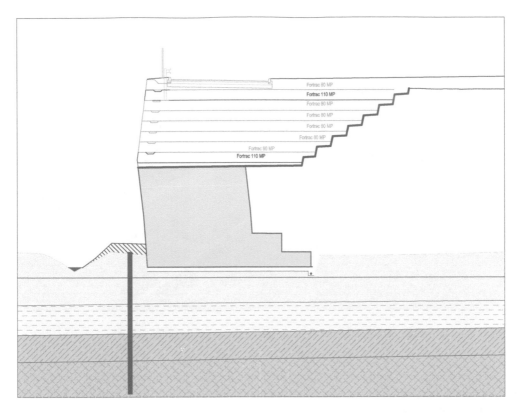

Fortrac 80 MP
Fortrac 110 MP
Fortrac 80 MP
Fortrac 80 MP
Fortrac 90 MP
Fortrac 80 MP
Fortrac 90 MP
Fortrac 110 MP

Figure 4.9 (Continued)

Figure 4.10 Installation of H piles at the wall toe to protect horizontal extrusion

Figure 4.11 Frontage roadway situated close to the reconstructed reinforced wall

The reconstructed wall was well monitored, and the final conclusion was that the frontage roadway situated close to the site perimeter and loading reconstructed wall could be fully loaded, Figure 4.11.

When summarizing the described case, a short conclusion is important:

- Instead of ULS of the GEO type (overall and internal stability, bearing capacity and sliding), the ULS of the type EQU should be also evaluated;
- SLS – differential settlement should be evaluated, because there is a natural differential settlement easily explained by the difference between 1D and 3D settlements;
- Functional drainage system should guarantee that no water pressure is influencing the stability;
- Monitoring also plays a very important role, not only to indicate a problem, but also to prove that the wall is stable and fully functional.

Note 1: More examples of a combination of geosynthetics reinforcement with nails or anchors are presented in section 4.2.3.

Note 2: Geosynthetics soil reinforcement is a very effective method for the construction of barriers along a transport corridor. Steep reinforced fills serve as acoustic and light shielding.

4.2.2   Soil nailing

Soil nailing is generally classified as a method of soil improvement in cutting, during which a quasi-homogeneous body is created and operates as gravity retaining wall. The application of soil nails, which are generally steel bars or pipes, consists of inserting them into the drill hole and grouting them, or driving them into the ground using different technologies. To

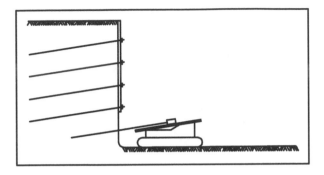 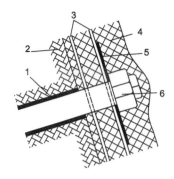

Figure 4.12 (a) Gradual insertion of nails, (b) detail of nail fixing 1 – cement grout, 2 – soil, 3 – steel mesh, 4 – shotcrete, 5 – head spreading plate, 6 – head of nail with nut, 7 – drainage mattress

guarantee the role of a quasi-homogeneous body, the number of nails should be at least one nail per 6 m². The length of grouted nails is about 0.8–1.2 times that of H (height of wall), and their capacity is about 200–600 kN. Usually nails are situated sub-horizontally (Figure 4.12a). Before the next row of nails is inserted, the wall face is cleaned. Subsequently, reinforcing steel mesh is fixed and shotcrete applied. A typical detail of the face of the protected wall is illustrated in Figure 4.12b.

Soil nailing technology is not utilisable for coarse soils with groundwater table above the toe of slope. Special care should be devoted to the drainage behind the wall face for fine soils, e.g. by applying toe drains or a permeable geo-matt. A thin gabion wall is able to fulfil both roles – to drain the face and at the same time to protect it, so that the shotcrete is not needed.

The first experiences with soil nailing were evaluated by Stocker (1979, 1994). Great progress was achieved during the French project "Clouterre" (Schlosser, 1991). Similarly, in the UK detailed information on soil nailing can be found in CIRIA C637 (Phear *et al.*, 2005). *CENTC 288 Execution of special geotechnical works* prepared the standard EN 14490 Soil nailing.

With respect to the soil nailing design, different approaches can be selected:

- *Nomograms* published by Schlosser (1991, 1993) are appropriate for the preliminary design;
- *Analytical calculation models* are similarly used as for a gravity retaining wall – stability along the slip surface passing behind the zone of reinforcement, safety against overturning or sliding, respective bearing capacity. They are supplemented by verification that the steel bar (tube) cross section is able to transfer tensile loading, and respectively that its shear strength is able to guarantee slope stability along the potential slip surface crossing the zone of reinforcement, Figure 4.13. More details are also described by Johnson and Card (1998).
- *Numerical calculation models* deal with the same problems as for geosynthetics reinforcement. A small difference is noted in modelling between nails and geosynthetics, as not only axial stiffness but also bending stiffness can be defined for nails. Numerical models are often preferred where precise deformation analysis checking SLS has to be performed. Simulation of the soil nail wall behaviour (deformation) for individual construction steps is also very useful. Predicted (modelled) values of displacements can be compared with measured ones, and in this way the design can be corrected when necessary. A similar practice is valid for tensile forces in instrumented nails.

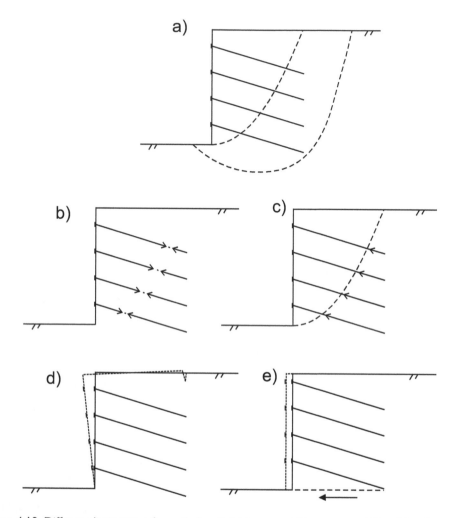

Figure 4.13 Different limit states for nailed wall: (a) inner stability, outer stability, (b) nail rupture, (c) nail pull-out, (d) wall overturning and (e) wall sliding

4.2.3   Combination of geosynthetic reinforcement with nailing

The combination of both reinforcement methods is typical for transport infrastructure widening or for landslide remediation.

Transport infrastructure widening is a typical solution to the growing demands on transport capacity. The basic problem is associated with new demands on land. The purchase of additional land around existing transport corridors is always a problem. In an effort to solve the problem and to save structural elements such as drainage systems or fencing, soil reinforcement offers a possible solution. The best possibility is in the case of inclined terrain, where reinforcement can help to construct steeper slopes, Figure 4.14.

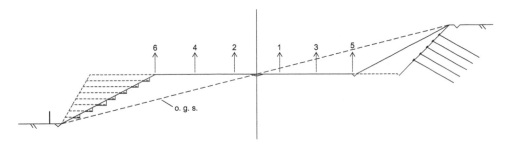

Figure 4.14 Cross section of widened infrastructure without new demands on land

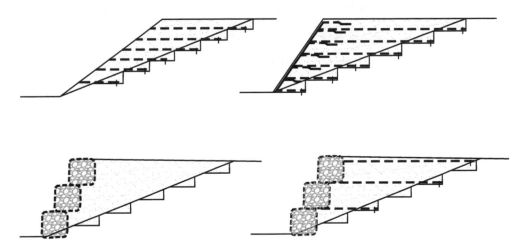

Figure 4.15 Embankment widening possibilities

More attention is focused on the embankment, where two points are sensitive:

- Contact of the existing slope with additional fill. Stepping of the existing slope enables gripping of the newly applied reinforcing element. This connection together with good compaction of new fill can eliminate the appearance of longitudinal cracks on the carriageway at the point of widening.
- Protection of the steeper slope against surface erosion.

For the case of relatively good subsoil, possible alternatives are illustrated in Figure 4.15. They encompass the case of wrapped reinforcement or the application of gabions. For nearly vertical slopes, concrete facing elements are preferred, either small or large ones (Figure 4.16).

For the case of weaker subsoil, the solution is more complicated, as was shown in Chapter 3. Problems are associated with bearing capacity, differential settlement and horizontal deformation. Kempfert *et al.* (1997) described an interesting solution to such a case. They used geotextile-encased sand columns in soft ground to eliminate the aforementioned problems.

Figure 4.16 Reinforced retaining walls with concrete facing prefabricates, small blocks and large panels

One case of slope remediation with combined reinforcement was described in detail by Vaníček (2017). A landslide occurred on an excavated slope in natural ground, on top of which a roadway was situated. The progressively collapsing slope was up to 8 m in height with a slope as steep as 1:1. The landslide needed remediation, as it created a relatively high risk for the roadway as well as for a proposed housing development project, situated close to the slope toe. This development project was situated on brownfields, in an area of former brickworks.

Geotechnical investigation rated brick clays as CS material – sandy clay firm to stiff consistency. Bedrock composed from weathered clayey shales was found at a depth close to 4 m in the upper part of the slope up to only 1 m in its lower part.

Figure 4.17 shows a cross section of the proposed and subsequently realized solution. Due to limited space, the slope had to be terminated by a retaining wall with a maximum height of 1.6 m. Based on the main architectural requirement, the retaining wall should have gabion facing. At the top edge of the slope, it was not possible to do any large excavations due to the aforementioned roadway. Driven soil nails both guarantee fixation of the geosynthetic reinforcement and cut through the potential slip surface, guaranteeing better connection with the new slope.

An extreme case of slope reinforcement was described by Detert and Fantini (2017). They describe the Trieben-Sunk project in Austria, construction of new roadway B 114 between the cities of Trieben and Judenburg. An old existing road displayed many problems due to permanent slope movement and damage to the retaining structures, a rupture of anchors to back-anchor bridges. These problems required cost-intensive maintenance work, e.g. the asphalt layer reached in an extreme place a thickness of up to 2 m due to settlement compensation. Trafficability of the road during winter was very dangerous, also as the result of high inclination – up to 20%. Due to previous experience and with respect to difficult ground conditions (steep terrain, landslide-prone area, creep deformation), a flexible solution was preferred, consisting of:

- Reduction of road inclination by serpentines;
- Geotextile reinforced embankments, which are able to compensate deformation to a certain extent without damage, whereas the height was up to 28 m.

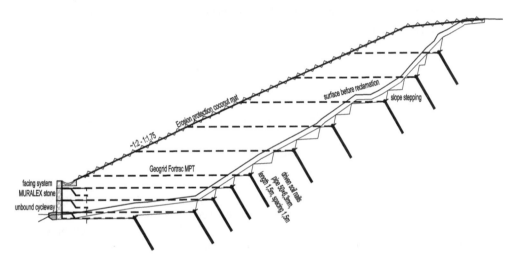

Figure 4.17 Application of both reinforcement methods for reclamation of collapsed slope.

Cut slopes were stabilized by anchors and their surfaces protected by shotcrete. The inclination of the reinforced fill, adjoining the steep rock slope, was about 60°. Stability was controlled by analytical as well as numerical calculation models. Final success was supported by:

- Construction of a stable base at the bottom of reinforced fill;
- Placement of lost formwork on the slope face, guaranteeing good compaction as well as erosion protection;
- Connection of geosynthetics reinforcing layers to rock slope by anchors, protecting the fill against tilting;
- Drainage of the water-bearing potential slip plane.

Figure 4.18 shows the final arrangement of the road serpentines.

Figure 4.18 Final arrangement of road slopes – Trieben-Sunk Project, Austria
Source: Courtesy of O. Detert.

4.3 Natural aggregate savings

There are roughly two basic possibilities for decreasing the consumption of natural aggregates:

- Improve the properties of natural aggregates;
- Apply nonstandard aggregates.

This section deals only with the second possibility, because the first possibility via soil reinforcement was already mentioned in the previous section, as it also leads to a decrease in land consumption. Soil improvement via the diffusive method by soil stabilization will be discussed in the next section, as it is also able to save energy.

4.3.1 Application of nonstandard aggregates

Different types of nonstandard aggregates are now available which fulfil the requirement of replacing large volumes of standard aggregates in earth structures. Such aggregates include:

- Excavated soil and rock obtained during the realization of foundation or underground structures; material excavated during metro construction is a typical example;
- Construction and demolition waste;
- Material obtained during mining of natural resources, as are different waste rocks. For the Czech Republic, clayey soils overlying brown coal seams are a typical example of this type of waste;
- Waste after the utilization of natural resources, as fly ash from electric power stations, slag from metallurgy and waste (tailings) from concentrator factories.

In all cases, efforts should be made to understand how the residues generated could be introduced in the revaluation chain, in which they could be labelled as raw materials.

Therefore, their utilization in earth structures for transport infrastructure is not only a possibility but probably the main purpose. Deciding which phase of the revaluation chain the particular residue can be used depends on two different validation processes:

- Validation of mechanical-physical properties, whereas testing methods are focused on structural stability, on sensitivity to structural collapse or on the swelling potential;
- Validation of chemical properties, especially chemical properties of leachate, to be able to qualify the potentially negative impact on the environment.

Note: The second process will be covered in section 4.3.1.6 for all alternative residues and nonstandard aggregates.

4.3.1.1   Construction sector – excavated soil, rock

Nowadays a great surplus of soil is being excavated in cities during construction sector activities. This is in agreement with the general tendency to utilize construction sites as much as possible. Underground garages are a typical example. However, the largest volumes of excavated soil are coming from underground critical infrastructure such as road and railways tunnels, metro systems, sewage systems etc. In most cases, the excavated material is natural, without any chemical contamination. Nevertheless there is ongoing discussion about whether we have to look at excavated material as waste material. It depends on legislation in the different countries. The attention in this respect is currently focused on quick and cheap methods of controlling possible soil contamination.

Excavated materials have different possible uses; however, with respect to volume and material differences, earth structures for transport engineering are the most typical. A fundamental condition for utilizing a large volume of nonstandard aggregates (first from the construction sector) is associated with the database of such aggregates (Vaníček *et al.*, 2017a). Such a database should include:

- Volume of waste (nonstandard aggregates), place of production and the place where such waste is stored;
- Declaration of the material properties, especially from the mechanical point of view;
- Declaration of the environmental properties – mainly character (quality) of leachate.

All this information creates a prerequisite for the potential opening of the commodity exchange, ensuring the balance between supply and demand (Gazda *et al.*, 2008).

With respect to the material differences, the first step before making a decision about using a particular material is its classification. With the help of up-to-date experiences (codes), the suitability of soil for earth structures for transport engineering can be evaluated. The attention should be focused on soft rocks and hard soils, such as claystone, clay of hard consistency, siltstones and clay shales. Macropores remain between individual grains after the compaction of such aggregates. As the contacts of individual grains for such aggregates are sensitive to softening, there is a potential risk of great additional settlement. Placement of such aggregates on a provisional deposit for a certain time can lead to an increase in moisture content. The resulting utilization can lead to better compaction with significant limitation of macropores. But an increase in moisture content should not exceed a certain limit, which would lead to soil degradation up to soft consistency.

4.3.1.2 Construction sector – construction and demolition waste

The volume of construction and demolition waste is increasing with time. Bricks and concrete crushed to a certain fraction create the main portion, with smaller amounts of ceramics, plaster or wood. Wood is undesirable as an aggregate, as are some other components, e.g. materials used as insulation. These soft or degradable components decrease material utilization and therefore should be sorted out. The commodity exchange provides an interesting opportunity for the effective utilization of construction and demolition waste.

In the Czech Republic, for example, the total amount of construction and demolition waste produced is about 3.5 million metric tonnes per year. This amount is disposed of at 160 operating recycling plants, either permanent or portable facilities, Figure 4.19.

When applied to the earth structures of the transport engineering, different gradings and strengths can be ordered according to the prevailing initial material – concrete or bricks. For earth structures, such recycled material can substitute for coarse soils – gravel or sandy gravel, either for fill or for improvement of the underlayment of new access paths, roads etc. For earth structures with low risk, untreated material – without sieving and separation into different fractions – can be used as well.

This waste material can hardly compete with natural sandy gravel used for concrete. However, we will mention one example of non-traditional application (Vodička *et al.*, 2008; Vaníček, 2011; Vaníček *et al.*, 2017a). In this case, crushed parts of old concrete and bricks were mixed not only with cement and water but also with short synthetic fibres. The final product after mixing created a new structural material, the properties of which can be different, Figure 4.20. They strongly depend on the amount of fibres and cement, as well as on the degree of compaction. Thus the different properties, either from the strength or respective permeability point of view, can be obtained.

A compacted layer of this brick-fibre-concrete, with a thickness of about 0.3 m, can create a reinforcing element in fill, similar to geosynthetics reinforcement. A special application will be presented in Chapter 5, where the resistance of the fill against surface erosion (during overflowing) will be discussed.

Figure 4.19 Recycling plant of construction and demolition waste in operation

Source: www.topdesign.cz/underwood/download/images/powerscreen.jpg (public domain).

Figure 4.20 New structural material – mixture of old pieces of bricks and concrete with short fibres and cement

4.3.1.3   Mining activity – overlying rock, waste rock

This chapter is focused on by-products during mining activity. For open pit mines, a typical example is overlying soil. For deep mines, either for coal or other raw materials, it is waste rock, which is the result of the separation of excavated material.

The first case will be described in more detail, as the authors have great experience with this by-product (e.g. Vaníček, 1986, 1995; Dykast, 1993; Dykast *et al.*, 2003; Vaníček and Vaníček, 2008). A coal seam is covered by tertiary sediments with a thickness up to 200 m in the north-western part of the Czech Republic. Claystone, clays of stiff to hard consistency, are predominant components of these tertiary sediments. A huge amount of this clayey material overlaying brown coal should be excavated and deposited, roughly 200 million m^3 per year. Some of the excavated material is deposited on the surrounding terrain in the form of so-called outer spoil heaps, and some is deposited back into the excavated pits, where they create inner spoil heaps.

The landscape of this region is extremely affected by mining activity on the one hand and by deposition of excavated material on the other. When the average height of these spoil heaps is about 40 m, they will cover 5 km^2 per year. Therefore, construction activity, including transport infrastructure, should use the surface of these spoil heaps.

Figure 4.21 (a) Final deposition by free fall and (b) character of individual clods after deposition and weathering

The most important problem with the surface utilization of spoil heaps consists in large settlement, with a limit state of serviceability. The huge bucket-wheel excavators are used for excavation. Very large individual clods sometimes should be crushed down, even when the width of a belt transporting them is between 2 and 3 m. Final deposition is by free fall from the overburden conveyor bridge from a height of roughly 20 m, Figure 4.21a. The character of such deposited material looks like rockfill, with a bulk density of approximately 1500–1600 kg/m³, Figure 4.21b.

The individual macropores between the individual clods are interconnected, as macro porosity is about 30%. Such material is permeable for air as well as for water. The process of particles degradation can start, also supported by the fact that inside of clay clods is negative pore pressure due to unloading. Under this condition, such soil will easily absorb water – free water or condensate water from saturated air – even from the air inside of the spoil heap's body.

The properties of the deposited clayey soils change with time due to two basic contradicting aspects (Figure 4.22):

•   Process of softening as a result of weathering – moisture content increase and kneading;
•   Process of hardening as a result of surcharge by new deposited layers.

It is obvious that engineering efforts support the process of hardening by limiting access of water inside the spoil heap body.

Two examples were well documented for a mining area in North Bohemia:

•   Construction of motorway D8 on a backfilled pit (Kurka and Novotná, 2003) describing geotechnical problems of foundation of the bridges and embankments on man-made ground.
•   Construction of a high embankment composed of deposited clay clods (Dykast *et al.*, 2003), (Vaníček and Vaníček, 2008) on which the crest of the transport infrastructure was situated – motorway and railway, together with four pipelines for diversion of the river Bílina.

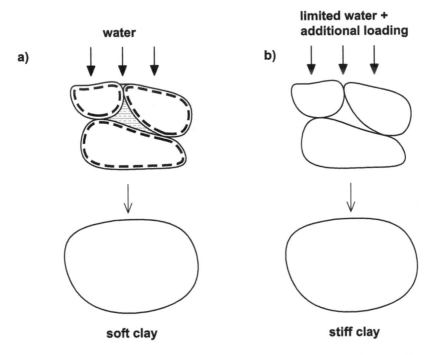

Figure 4.22 Different changes of clay clod properties for different possibilities of water infiltration

The second case, reported as the Ervěnice transport corridor, will be described in more detail. This inner spoil heap has been filled between the Šverma and ČSA mines belonging to the North Bohemian brown coal basin. As the average height is about 130 m, with a slope inclination of roughly 1:7 and a length of about 3.6 km, this corridor was at the time of construction the largest earth structure of transport infrastructure in the world, Figure 4.23. A basic question from the early stages of the project was associated with settlement prognosis over time.

Filling of the corridor ended in 1983. The attention was focused not only on the maximal value of the corridor surface settlement but also on the differential settlement in longitudinal axis (Dykast, 1993). The prognosis was based on two types of information (Vaníček, 1991):

- Up-to-date experience with well-documented existing spoil heaps;
- Lab experiments utilizing a large diameter oedometer, enabling to test large clods with different initial moisture content.

Subsequent corridor monitoring approved that the prognosis was in very good agreement with measured values, with a total surface settlement of close to 2 metres. Great discussion at the end of filling was concentrated on the arrangement of the last metres, not only to eliminate total settlement (as uncompacted fill still has high macroporosity) but also to eliminate differential settlement. Due to the pressure on the corridor opening only a segregation of fill was applied there, where soils not as sensitive to degradation were preferred. The technical recommendation to postpone the corridor opening was not fully accepted, even when all partners were informed that the rate of settlement is significantly decreasing with time.

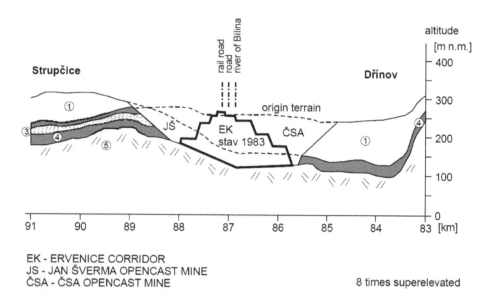

EK - ERVENICE CORRIDOR
JS - JAN ŠVERMA OPENCAST MINE
ČSA - ČSA OPENCAST MINE   8 times superelevated

Figure 4.23 Scheme of the Ervěnice corridor: I – overlying clays, 2 – upper coal layer, 3 – sandy clays and sands, 4 – main coal layer, 5 – subjacent rocks

Figure 4.24 Surface of the spoil heap utilization for transport infrastructure

Therefore, the speed limit for trains was limited at the beginning, roughly to 40 km/hour; however, within few years it was possible to start regular operation, Figure 4.24.

With time, another question must be asked – what will be the impact of backfilling the remaining parts of the excavated pit, which will cause additional loading on both slopes of the corridor embankment?

4.3.1.4   Industrial waste – ash, slag

Industry is producing an extremely large amount of residues, which can be denoted as a waste. To decrease land consumption for different landfills or tailing dams, where these waste materials would normally be stored, engineered effort is directed at different possibilities of utilizing waste. Waste utilization could be either in produced form or recycled form, which would guarantee much better mechanical and environmental properties. With respect to the utilization for earth structures, two such products will be mentioned – ash and slag.

ASH

Burning of solid fuel for production of electricity and heat is still the prevailing manner of energy production in the world, which is also roughly true for the Czech Republic. Therefore, the production of sediments from solid fuel burning is high, reaching about 1.1 tonnes per habitant per year. This is due to the high content of non-burning particles in brown coal and also to high energy demands generally. The utilization of ash in earth structures for transport engineering is one of the most important ways to decrease the volume of ash deposited in tailing dams.

Properties of ash vary according to the type of coal, type of burning and type of desulphurization methods applied. For example, the utilization of the remaining parts after brown coal burning is more problematic than for black coal, due to the low content of basic compounds CaO and a higher content of acid compounds – Al_2O_3 and SiO_2.

From the geotechnical point of view, the particle size distribution curve roughly corresponds to sand and silt, the character is non-plastic and the shear strength is characteristic, with an angle of internal friction of roughly 35°, the value typical for sand. The filtration coefficient is mostly in the rage of 1×10^{-6} to 1×10^{-7} m/s, so it means that ash is relatively permeable. The main difference between ash and fine sand is in dry volume density, which is for ash very often in the range of 950–1050 kg/m³. On the one hand, it is an advantage, as embankment composed from ash can reduce settlement of subsoil. On the other hand, it is a disadvantage for slope stability when water seeps through the embankment body, Figure 4.25. If:

$$F_{geo} = (1 - \gamma_w/\gamma_{sat}) \, \text{tg} \, \varphi_d/\text{tg} \, \beta$$

When: γ_{sat} is ash saturated density ($\sim$14 kN/m³) and φ_d design angle of internal friction. When tg φ_d = tg φ_k / $\gamma_{m, \varphi}$. For $\gamma_{m, \varphi}$ = 1.25:

$$\text{tg} \, \beta = 1/6.25$$

which is much smaller than for classical sand (1:3–1:4).

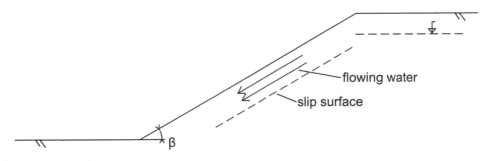

Figure 4.25 Scheme for the general slope inclination deduction for ash when water is seeping through it

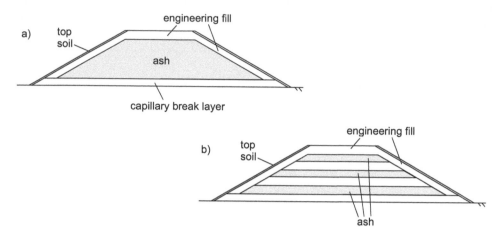

Figure 4.26 Different arrangement of ash in earth structure

Also, sensitivity to surface erodibility (elicited by either wind or water) is higher. Therefore, it is recommended to apply ash into the central part of cross section, Figure 4.26. The sandwich arrangement is preferred when the basic soil is of a clayey type, as more permeable ash can significantly speed up clay layer consolidation.

As was mentioned already, the properties of ash are also the result of desulphurization systems, namely when burning brown coal. Better properties have ash from fluid burning, when fine-milled coal is mixed with fine-milled limestone and this mixture is burned together. Better self-hardening properties predestine this type of ash for direct application.

Sometimes there is a problem associated with determining optimal moisture content, as for ash the character of its Proctor curve, its peak, is not as obvious as for classical soil. Another desulphurization system, called the wet lime washing method, is preferred for bigger electric units. Combustion gas is mixed there with water and with a mixture of fine milled limestone.

Note: One by-product of this method – energy-gypsum – is used for manufacturing plasterboard construction materials.

The remaining ash is used after small modification with some additives as stabilized ash or reinforced ash. Figure 4.27 shows the application of stabilized ash in the "wet" form.

Nevertheless, the final product is rather brittle and susceptible to crack formation. To improve its flexibility, new experiments with ash reinforced by micro synthetic fibres have begun, and they are concentrated not only on mechanical properties but also on the technology of mixing (Vaníček and Jirásko, 2017), Figure 4.28.

SLAG

Slag is a by-product of different thermic and burning processes. Metallurgic slag is the result of smelting and refining metals. A simplest division is on blast-furnace slag and steel slag, both rising as a by-product during conversion of iron into steel.

As the character of slag is close to gravel, there is a tendency to use slag as a substitute. The subgrade of large hall floors or asphalt motorways are typical examples where slag can be substituted for gravel. However, attention should be devoted to the relatively high

Figure 4.27 Deposition of stabilized ash in the "wet" form for the road foundation
Source: Courtesy of S. Chamra.

Figure 4.28 Lab tests of composite of ash with short synthetic fibres

Figure 4.29 Curly surface of motorway D1 close to Ostrava where slag sensitive to swelling was used

Source: www.idnes.cz/ekonomika/doprava/skody-za-zvlnenou-dalnici-jdou-do-stamilionu-rsd-proveri-dalsi-useky. A120102_171819_eko-doprava_vem.

swelling potential of some kinds of slag. Figure 4.29 shows the curly surface of motorway D1 close to Ostrava, where slag sensitive to swelling was used.

From this point of view, Motz and Geiseler (2000) describe recent experiences with applying steel slag in the construction sector in Germany. They distinguish between "basic oxygen furnace slags" (*BOF* slags) and "electric arc furnace slags" (*EAF* slags) as the result of different production processes. The basic difference is in the chemical composition, expressing whether dolomite or lime was used in the production process. *BOF* slags contain up to 10% of free CaO and up to 5%–8% of MgO, while *EAF* slags have less CaO and more MgO. The authors emphasize that the volume of free CaO and MgO is the most important factor with respect to volume stability. To evaluate the swelling potential, they recommend a steam test, which is now part of EN 1744–1. They stated that slags with a swelling potential up to 3.5% can be used for asphalt bonded mixtures and up to 5% for aggregates for unbonded mixtures. However, even slags with higher swelling potential can be used. A typical case is for anti-erosion protection of slopes for higher volume density of these aggregates. Other authors also summarize their experiences (e.g. Lind *et al.*, 2000, for Finland and Urbina *et al.*, 2000, for Spain).

4.3.1.5   Others

Other nonstandard aggregates can be used to fulfil specific roles. They are not typically used for the main part of earth structures. One such example is foam glass, also called foam glass granulate, made of recycled waste glass from small pieces of glass which are not appropriate for recycling in glass works. The final product is composed of differing grain sizes, e.g. 16–32 mm, is lightweight (free-fall density about 150 to 170 kg/m^3) and has very good insulation properties. As individual grains are shape-stable with a close inside structure, the

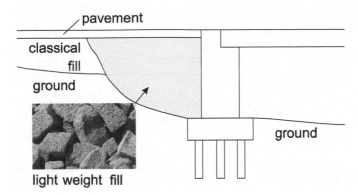

Figure 4.30 Character of foam glass and utilization on the contact of fill with bridge abutment

water infiltration into individual grains is practically zero. Only a small amount of water can infiltrate due to capillarity. The specific roles that can be served by such material include:

• Lightweight fill – to reduce ground settlement or to reduce earth pressure;
• Insulation;
• Drainage.

The combination of these specific roles can be appropriate for fills in contact with a bridge abutment, Figure 4.30. This backfill can be also reinforced, as was mentioned by Jelušič and Žlender (2019).

Note: Similar application as foam glass can be for lightweight aggregates from expanded clay.

Great care should be devoted to structural collapse, similarly as for other nonstandard aggregates. The resistivity against grain breakage (collapse of the unique enclosed structure) for foam glass is, according to one producer's data, roughly 1.24 N/mm².

4.3.1.6   Environmental aspect of nonstandard aggregates

As previously mentioned, the specific branch of environmental geotechnics covers all environmental problems associated with geotechnical engineering. The problem of contamination spreading in the ground creates a very important part of it. It is outside the scope of this publication to describe the problem from different aspects, as they are covered in many other publications (e.g. Young, 1992; Sutherson, 1997; Rowe, 1992; Vaníček, 2003; Jirásko & Vaníček, 2009).

In this section, we will concentrate only on the basic problems of the contaminant spreading modelling from the embankment containing nonstandard aggregates. Such modelling serves for the risk assessment of the environmental impact of this embankment type. A schematic illustration of the problem is presented in Figure 4.31.

CONTAMINANT SOURCE

In principle, the contaminants can be divided into organic or inorganic ones. Specific position between organic contaminants have contaminants of the liquid character, poorly dissoluble in water, denoted as $NAPL$ – "non-aqueous phase liquids". They can be subdivided into lighter than water ($LNAPL$) or denser than water ($DNAPL$). The most frequent inorganic

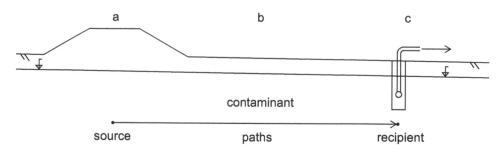

Figure 4.31 Schematic illustration of the contaminant spreading problem: (a) contaminant source, (b) contamination paths and (c) contaminant recipient

substances which lead to contamination are metals, cyanides and ammonia. From the metals, most problems are created by chromium, cadmium, zinc, lead, mercury and arsenic.

In the ground, organic contamination can in principle co-exist in four different phases as:

- Vapour in pores of soil;
- Independent transportable liquid phase;
- Adsorbing phase;
- Partly soluble in water, as soil humidity.

Primary disqualification of some nonstandard aggregates from their application is based on the comparison of the composition of chemical components with defined limits. Usually these limits are defined on national or international levels. Lower limits roughly correspond to the natural background. Significant risk for human health or environment in general is defined by exceeding their upper limits. The cases ranging between these limits deserves a risk assessment process.

CONTAMINATION PATH

The length and character of the contamination path have a significant impact on the change of the initial degree of contamination c_o to the final one c when reaching the contaminant recipient. Permeability has an important role, as do other characteristics influencing contaminant spreading. The level of groundwater and its fluctuation are also very important. Therefore, knowledge of the permeability and the interaction of a specific contaminant with soil minerals is a basic presumption.

CONTAMINANT RECIPIENT

The water source is the most significant recipient. A great distinction is made between sources of potable water and a water stream. In the second case, the degree of contamination is lowered by dilution, in an ideal case up to natural background.

In the risk assessment process, the equation of contaminant spreading plays an important role. For the 2D problem, this equation can be written in the form:

$$n\frac{\partial c}{\partial t} = n \cdot D_x \frac{\partial^2 c}{\partial x^2} + n \cdot D_z \frac{\partial^2 c}{\partial z^2} - n \cdot v_x \frac{\partial c}{\partial x} - n \cdot v_z \frac{\partial c}{\partial z} - \rho_d \cdot K \frac{\partial c}{\partial t} \qquad (4.7)$$

Which express that the contamination spreading depends more on transportation processes as:

- Advection;
- Diffusion;
- Dispersion;
- Sorption.

This equation is able to specify the change of contamination degree as a function of position x and z and time t, in dependence on:

v – velocity of groundwater flow;
D – coefficient of hydrodynamic dispersion.

Where $D = D_e + D_m$

When D_e – effective diffusive coefficient – influences the movement from the place of higher concentration to the place of lower concentration; and

D_m – coefficient of mechanical dispersion – influences the mixing and scattering of contaminants due to subsoil inhomogeneity.

K – distribution coefficient – relates to sorption, to the process decreasing contaminant spreading.

From the practical point of view, the applied nonstandard aggregates have to be deposited in an embankment with no direct contact with the groundwater table (even for the most dangerous scenario). A drainage layer is applied between the embankment and subsoil to limit capillarity as well. Only rainwater penetrating the embankment could leach contaminants from material used for the construction.

Finding a solution to contaminant spreading can be done by using either a deterministic approach for selected input data or a stochastic approach, using the range of values obtained during laboratory tests for individual input data (Vaníček, 2003). For the second case, the solution is mainly based on the Monte Carlo method, which means that the model is analysed several thousand times with randomly generated input parameters that satisfy the entered probability distributions. The output shows probability exceeding given trigger values at a given time. Vaníček (2006) describes such an approach applied to the practical problem of a transport embankment utilizing ash.

4.4 Energy savings

Energy savings is another important aspect of the sustainability approach. The pressure on energy savings originates from many different directions, for example:

- From environmentalists – with the main aim to protect natural resources for future more effective utilization or just to decrease the carbon footprint to have some impact on climatic changes;
- From economists – as the cost of energy increases because of higher demands on it;
- From politicians – to decrease dependency of the countries with lower reserves of energy sources on the countries with higher sources.

As transport activities generally consume a relatively high percentage of total energy (roughly about 30%–40%), the pressure on energy savings is also very high there.

Also in the field of earth structures for transport engineering, some possibilities are discussed and applied, such as:

- Savings via lower demands on excavation and transport;
- Savings during the application of aggregates into the earth structure, above all during soil compaction;
- Savings via the application of new smart geotechnical structures;
- Savings via geothermal energy utilization.

All of these possibilities will be discussed in the following sections.

4.4.1   Soil stabilization

The most frequent type of soil excluded from direct utilization in earth structures for transport engineering are fine (clayey) soils with a moisture content higher than optimum. The excavation of such soil from the part of infrastructure corridor – which is in cuts, with its deposition outside of direct utilization, followed by excavation and transportation of more appropriate soil from greater distances, results in higher energy consumption. Soil stabilization of the fine (clayey) soil, firstly by lime, is a good solution for such a situation, Figure 4.32.

Note: There are other ways to stabilize soil, but they are more typical for stabilization of base courses or are used for temporary stabilization of a road surface, e.g. mechanically stabilized soil using the mixture of two soil types. The aim is to reach such composition that coarser grains will form the skeleton and finer grains will fill the pores between the coarser grains. The final product can form the best quality unbound base course of roads. Similarly, stabilization with the help of cement is preferred for coarse soils. However, other bonding agents are also used, such as fly ash, dust from rotary cement kilns, granulated blast-furnace clinker, slag from steel mills or combinations of these.

Figure 4.32 General view on earth cutter for lime stabilization

From the different types of lime available, quick lime is preferred. The amount of quick lime is roughly in the range of 1% to 6%. Percentage is influenced by the main aim of the application:

- Workability improvement;
- Strengthening improvement.

Hydration starts when the wet soil is mixed with quick lime. The result of this exothermic reaction is slaked lime. The heat generation has an immediate effect, which is attributed to the better compaction of these wet soils. Therefore, it is necessary to distinguish between the immediate (short-term) effect, connected with better compaction, and the long-term effect. While the short-term effect occurs always, the long-term effect will occur only for higher quick lime content. Roughly, this limit is close to 2% of lime to dry soil. For higher content, a character of the grain-size distribution curve is changed, as the result of particle flocculation. Treated soil creates lumps and diverse structure. Plasticity limit w_p is generally increasing, with minimal change in the liquid limit w_L, decreasing by this way the plasticity index I_p.

From the long-term point of view of improving the geotechnical parameters, the lime percentage should be roughly in the range of 2% to 6%. The development of cementitious bonds is attributed to the fact that clay is a natural pozzolana containing silica and alumina. Formation of cementitious bonds therefore requires a minimum amount of clay minerals, generally evaluated in the range of 10% to 17%. A long-term stabilization effect leads to:

- Strength increase with higher bearing capacity of the ground; however, the creation of cementitious bonds is a function of time and therefore strength increase should be tested with some delay, at least after 28 days;
- Reduction of the susceptibility to swelling respective shrinkage, generally to the structural changes;
- Reduction of moisture content, resulting in better workability and more effective compaction. Moisture content reduction is the cause of more aspects. Moisture content is first reduced by hydration heat and by absorption during the change of lime from quick to slaked. At the same time, the reaction of lime with clay minerals decreases plasticity, and this influences another factor relevant to better workability.

From this statement, one would gather that lime stabilization can solve most of the problems connected with the utilization of clayey soils. Some authors, however, such as Greaves (1996), Smith (1996), Brandl (1999) and Vrbová (2008), stress some factors to which great care should be devoted.

a)   Soil stabilization is typical for earth structures connected at least with small risk. Therefore, there is a preference for one's own lab tests, devoted not only to the workability as optimum water content and maximum dry density, but mainly to determine the geotechnical data needed for the calculation model for the limit states analysis. Usually, a percentage of lime higher than 6% has no additional effect on strength increase. For very important projects, the compaction trial is worthwhile.

b)   The mix-in-place method of the manufacturing procedure is now preferred, but the final result is sensitive to the individual steps. Soil prepared for stabilization is first modified, then the bonding agent is spread all over with the batcher and successively mixed with soil using the cutter. If there is such possibility, between mixing lime with soil and final compaction there should be some time delay, roughly 24–72 hours. This time delay is denoted as mellowing or maturing. Following each mixing pass, the material is trimmed and lightly compacted. Sheep-foot rollers can be used for thicker layers, but they must be followed by a smooth roller.

c)   Non-success with the lime stabilization was registered in cases where the soil or pore water contained sulphates, where sulphates arose through oxidation of "sulphides," or where the soil contained a large amount of organic matter. In these cases, the reaction with lime causes swelling and a successive reduction in strength. This swelling is attributed to the creation of the mineral ettringite, for which hydration is accompanied by large dimensional changes.

d)   Combination of lime with cement. Lime prepares soil for cement (as is changing its typical clay character) and should be added at least some hours before the cement is added. This combination is especially recommended if heavy clays should be stabilized with lime during late autumn with potential exposure to frost.

4.4.2   Energy savings via intelligent compaction

During soil compaction, a significant amount of energy is consumed as well as during subsequent compaction control. Therefore, knowledge of the different factors influencing the final compaction result is very useful. If we suppose that the moisture content of soil prepared for compaction is in the range of acceptable limits, we can then speak about the following factors:

* Thickness of compacted layer;
* Frequency and amplitude for vibratory techniques;
* Type of compaction technique.

For the earth structures for transport engineering, there is a general tendency to select a higher thickness of compacted layer, approximately in the range of 0.3–0.6 m. To guarantee good interconnection of neighbouring layers as well as good compaction of the lower part of the compacted layer, the roller weight (or the weight per unit of roller axis) should be able to fulfil these demands. The number of roller passes to reach the demanded dry density at the bottom of the compacted layer is an important piece of information, Figure 4.33. Bigger construction companies with a large range of rollers can select the most optimal one with lower energy demands.

Nowadays, vibratory rollers dominate. Of the two basic vibration components, usually amplitude has a greater impact, as the kinetic energy of the vibrating mass is transferred into supplemental static pressure. However, there are some limits with respect to the health of the workers operating the roller or to roller deterioration. An ideal frequency corresponds to the resonance frequency of the compacted soil and the compacting roller. This ideal frequency is roughly in the range of 800–1700 oscillations per minute (about 13–28 Hz), depending on the soil type, its compaction rate and characteristics of the compacting roller used. The impact of these two basic vibration components is seen in Figure 4.34, when scanning the pressure at the base of the same initially compacted layer. To ensure a sufficient amount of oscillation impacts on area, the roller speed is relatively slow – 1.5 to 6.0 km/hour.

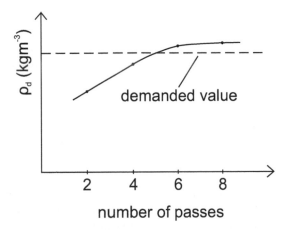

Figure 4.33 Dependence of final dry density on the number of roller passes

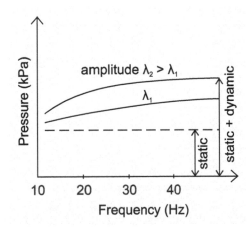

Figure 4.34 The impact of two basic vibration components on total pressure in a compacted layer

As mentioned previously, the heavy vibratory rollers are preferred for main parts of the earth structures. Smaller rollers or vibrating plates and rammers are typical for contacts of the earth structure's body with some other structures, where there is a tendency to reduce a horizontal component of the compaction pressure. Typical cases are bridge abutments, contact with different subways, face bracing of steeper slopes etc. Single vibratory rollers can differ:

- From the aspect of its surface – smooth, rubber-tyre or sheep-foot. Apart from these possibilities, relatively unusual shapes are becoming available. Adam and Markiewicz (2001) describes a polygon drum, which is composed of five octagonal elements. Similarly, Pinard (2001) describes rollers with non-circular compacting masses – termed as high-energy impact compactors.

- From the aspect of vibrating effect application –Brandl (2001b) mentions classical vibratory rollers, oscillatory rollers, Vario rollers and Vario Control rollers. While a vibratory roller exerts mainly vertical surcharge by a rotating mass excentre (tapet), the drum of an oscillatory roller oscillates torsionally. In a Vario roller, two counter-rotating exciter masses, which are concentrically distributed on the axis of the drum, cause a direct vibration.

Vibratory rollers denoted now as rollers with continuous compaction control (CCC) or as intelligent compaction, are able to guarantee basic conditions described in previous points, but the main source of energy savings is via two points:

- Saving energy via optimization of the number of roller passes. For example, for classical rollers an increase of the number of passes from six to eight represents an energy increase of about 33%. This increase is the result of some uncertainties or as the result of concerns about possible insufficient compaction. The main advantage of the CCC method is to limit subsequent additional compaction. The method is based on the interaction between the acceleration characteristic of the vibrating roller drum and the stiffness of the soil changing with progressing compaction. For validated conditions, the driver is informed when the demanded stiffness is reached. Obtained results lead to the optimization of the number of passes and are at the same time a guarantee of sufficient and uniform compaction.
- Saving energy needed for compaction control on collected samples in the lab. This is only an additional aspect from the perspective of saving energy, as the guarantee of work continuity is the main factor there.

4.4.3   Geothermal energy

Interest in renewable energy has been quickly rising during the last few decades. Geotechnical engineers are focusing on geothermal energy; for example, under the umbrella of ISSMGE there is a technical committee devoted to this problem. Also, there are common European research projects, e.g. COST project called GABI.

Geothermal energy utilizes the heat energy stored in the centre of the earth; the energy utilisation is increasing with the depth from the surface. The utilization of geothermal energy from small ground depths is possible through the contact of this ground with pipe filled with a moving medium, mainly water. In the case of so-called thermo-active structures – energy piles (but also diaphragm walls or tunnel lining), the pipe is connected to the armature of these piles. With the help of reversible heat pumps, the ground can be used as the source of heat energy (during winter) or as heat storage (during summer).

As piles are not typically used for the earth structures for transport engineering, with the small exception of a piled embankment, attention is focused on bridge foundations – piers, abutments.

However, an important question is where to use this energy. One possibility is to heat the transport infrastructure surface, mainly to protect it from freezing during the winter. The first studies on this topic show that surface heating is a realistic option. However, the amount of heat extraction from energy foundations is relatively low compared to the demands on energy to significantly decrease the temperature range changes in concrete of an integral bridge (Sehnalová, 2018).

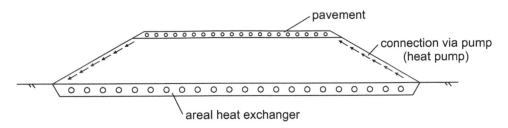

Figure 4.35 Earth aerial heat exchanger for heating and cooling of the motorway surface

With respect to the up-to-date results for surface heating, our attention is now focused on the earth areal heat exchanger. The most appropriate place to install such an exchanger is where the embankment (preferably high) makes contact with the ground, Figure 4.35. This possibility offers different options for a final solution.

A numerical model is needed to design the thermo-active system and determine its efficiency, including the calculation of the energy yield.

The equation of heat spreading in soil can be written in the form:

$$\frac{\partial T}{\partial t} = a\left(\frac{\partial^2 T}{\partial x^2} + \frac{\partial^2 T}{\partial y^2} + \frac{\partial^2 T}{\partial z^2}\right) + \frac{Q_{source}}{\rho c} \qquad (4.8)$$

Where:
cT/ct is the rate of change of temperature over time at a point;
T is temperature as a function of space and time $T = T(x,z,y,t)$;
ρ is the volumetric mass density of subsoil;
Q_{source} is the heat extracted or stored by a thermo-active structure;
α is thermal diffusivity (m².s⁻¹) $\alpha = \lambda/(\rho c)$;
λ is the coefficient of thermal conductivity;
c is specific heat capacity.

The last two parameters that enter the calculation relate to the soil/rock environment.

The thermal conductivity coefficient λ (W/m/K) expresses the ability of a given substance to conduct heat. Its numerical value expresses the amount of heat that, at steady state, passes per unit of time through a cross section of the substance at a unit temperature gradient. The value of coefficient λ is not a constant parameter for a single soil/rock type. In general, the thermal conductivity value increases with increasing temperature. In addition to the temperature dependency, the conductivity coefficient value is also affected by the degree of soil saturation, dry bulk density, mineralogical composition and grain-size distribution.

The second basic thermal parameter of the rock is the specific heat capacity c (J/kg/K). This parameter expresses how much heat is to be added to increase the temperature of one kilogram of substance by one temperature degree. As with the thermal conductivity coefficient, the specific heat capacity value has an increasing tendency with increasing temperature.

4.4.4 *Smart geotechnical structures*

The term "smart" is very trendy now and is used for different applications. For the construction sector, the term "smart buildings" is typically used, but the term has two basic meanings: one is that such a structure requires less energy and therefore has a lower CO_2 footprint, and the other is in respect to services, how the operation of the building is directed by IT technologies.

For us the first term is most important. Chang *et al.* (2019) quote the report of Intergovernmental Panel on Climate Changes that the atmospheric concentration of CO_2 rose from 280 ppm (mg kg^{-1}) in 1750 to 400 ppm in 2015, which means increase of about 42%. Smart structures should at least try to stop this increase, thereby adopting the Paris Agreement, which was signed by 197 countries in 2015. For the construction sector, it is important to know that cement-related CO_2 is reaching 10% of all production of CO_2 and has roughly doubled during the last four decades – about one ton of CO_2 is generated to produce one ton of cement.

Therefore, beyond the already mentioned possibilities of how to decrease energy consumption, more of our efforts should be focused on concrete structures such as piles, retaining walls, bridge abutments etc. One way is to find a substitution for cement, and the second is to prove that energy consumption for the alternative solution will be lower over the expected life span of the structure. This should include not only the period of construction (when the energy demand of individual materials is most important), but also during its service life as well as its dismantling.

Reinforced earth structures are a typical case of such substitution, especially reinforced retaining walls and soil reinforced bridge abutment. A good example in this respect was presented by Heerten *et al.* (2013), based on experiences in Germany. A classical cantilever concrete retaining wall was substituted by a revegetated steep slope (60°) from reinforced soil. The price was 1.6 times lower, but more importantly the CO_2 footprint was about 5–6 times lower. For a concrete wall, it would have been about 542 t of CO_2, while it was only 101 tonnes for the reinforced wall.

For carbon footprint comparison, the following individual steps were included:

- Cantilever concrete wall: production of concrete B 35; production of steel ST 500/550; transport of concrete and steel; installation; soil transport and backfill compaction;
- Geogrid – reinforced steep slope: production of geogrid; production of steel elements; installation; concrete beam; brickwork; soil transport and backfill compaction.

A similar study comparing geosynthetic-reinforced soil bridge abutment to concrete abutments is provided by Bizjak and Lenart (2018).

Such evaluation is not typical for geotechnical engineers, however to make this sustainable approach much more relevant, collecting information about energy demands for individual construction materials and works is needed.

There are currently a number of freeware and commercial computing tools for assessing CO_2 emissions for civil engineering. These programs include information on CO_2 emissions for individual construction materials, construction techniques and transport.

The following calculators that can be used for transport infrastructure construction have been selected as examples:

- Environment Agency carbon planning tool (e.g. gov.uk)
- Carbon emissions calculation tool: Highways England (e.g. gov.uk)
- Asphalt pavement embodied carbon tool (e.g. trl.co.uk)
- Carbon calculator of construction and demolition waste (e.g. wrap.co.uk)

4.5   Structure maintenance

Soil properties inside of an embankment body have a general tendency to improve with time. This improvement is attributed to the strengthening of bonds between individual particles as cementation or by so-called cold welding. This aspect is important from two basic points:

- Old embankment can be stable with limited deformation even when transport intensity is slowly increasing;
- From the maintenance point of view, the surface parts have the highest sensitivity to degradation.

The first point is valid for a slow process of growing transport intensity. However, for jump growth, e.g. during the reconstruction of a railway track for higher speed, for a higher number of trains per hour or for heavier loading, some measures should be taken. They consist (out of superstructure reconstruction) mostly of improving the contact between superstructure and substructure. Practically all the aforementioned methods of improvement can help. Starting from the substructure surface re-compaction, via substitution of the upper part of substructure by better material (e.g. gravel), respectively via their stabilization and finishing with geosynthetics reinforcement of this contact.

The authors have encountered another practical problem, attributed to the bonding forces. During the reconstruction of old bridges on old railway track, part of the embankment close to the bridge abutment was also removed. When an excavated material was redeposited on the previous place with the same slope inclination, the first symptoms of elevated deformation were observed when operation on this track started, even with the same transport intensity as before the reconstruction. But this intensity is higher than it was when the track opened more than 100 years ago. So this means that during excavation and re-deposition, the bonds between the particles were destroyed.

The second point connected with lower demands on the maintenance of earth structures should be taken into account before a final decision is made of whether to construct classical earth structures (embankments or cuts) or bridges, respective tunnels. It must be asked, for which height of embankment or depth of cutting is a classical earth structure more appropriate than the construction of bridge or tunnel? This economic question – the costs at the end of construction plus those for service time maintenance – is one of the most important. In this respect, Brandl (2001b) mentions the profitability of the additional embankment construction to the existing natural slope, on which crest the motorway will be running. Brandl states that modern compaction equipment, optimization and control have opened the possibility to construct high embankments instead of bridges for roads, highways and railways. With reference to financial demands, demands on maintenance and also to the more appropriate integration of green slope into the countryside, the alternative of high embankment (100–120 m) obtained the priority.

4.5.1   Slope maintenance

The slopes of earth structures require continuous servicing over their entire expected life spans. Surface erosion and weathering of cutting slopes are the main problems for slopes without any improvements. Surface water erosion depends upon:

- Resistivity of soil to water erosion – when for non-cohesive soils most sensitive to erosion are fine sands and silts and for fine soils most sensitive are dispersive soils.

- Intensity of the water stream – when the speed of surface water run-off depends on angle and height of slope, respectively, on the possibility of the creation of concentrated run-off.

Natural materials such as soft rock, e.g. claystone, are susceptible to weathering. Unloading due to excavation leads to swelling, and the subsequent freezing–thawing cycles can degrade ground surface into depth influenced by these cycles, typically to the depth within a range of 0.8 m to 1.2 m. Contact of weathered (softer) materials with non-weathered ones create a potential slip surface of shallow areal landslides. This process also applies to material of fill, though to a lesser extent.

Typical countermeasures are:

- Prevention against concentrated run-off – ditch in the upper part of slopes with safe outflow out of slope surface. Higher slopes should be stepped to reduce the speed of surface water.
- Resistivity increase by slope revegetation. For slopes of an embankment, revegetation with grass is the most used protection measure. The grass's root system connects the grass carpet with the subsoil for higher resistivity. Cazzuffi and Crippa (2005) tested different types of vegetation with longer and stronger roots, as they can decrease the potential of shallow landslide development.

As grass vegetation requires maintenance, shrubs or trees are preferred in upper parts of cuttings. Higher shrubs and trees have stronger roots, and they also can fulfil the protective function of the surrounding environment against noise and light from the transport infrastructure.

Geosynthetic grids or erosion protection mattresses can fulfil the erosion protection function until the green vegetation starts growing. Good connection of the green vegetation roots not only to the ground but also to this artificial protective system can increase resistance against surface erosion during heavy storms with extreme rain. Grids made from a natural degradable material such as jute or coconut fibre can be sufficient for short-term protection.

For steeper reinforced slopes, the protection function usually requires concrete facing elements – either small or large ones, mentioned already in section 4.2.3. The maintenance demands of such facing elements are very limited from the long-term point of view. The application of gabions with stone is another solution for steep slopes.

Maintaining the drainage system is very important, particularly for inner drains. Drain clogging can lead to an increase in the groundwater table, which has a negative impact on slope stability. Maintenance plans should take this fact very seriously.

Chapter 5

Availability and affordability approaches

Availability and affordability principles/approaches are getting increased attention lately on a general level – such as with the European project INTACT (Resilience of critical infrastructures in the context of changing climate and related weather events), respectively on more specific level as ELGIP Position Paper (2018). Some individual papers published inside of the geotechnical community can be mentioned as well, e.g. Rogbeck *et al.* (2013) and Lacasse (2013).

The main benefit of the available transport infrastructure is to guarantee secure and resilient solutions, mainly during natural and man-made hazards events. Briefly speaking, the availability aspect is becoming more significant with the higher demands on the mobility of people and goods. Good accessibility on the workplace, health care, recreation, people meetings respective delivery of goods or products on demanded place in specified time should be ensured regularly as well as during nonstandard situations created by both natural and man-made hazards (even on a limited scale). Therefore, strategic objectives should focus on developing innovative techniques and methods to reduce risk during disasters and adapt to climate changes.

Similarly, the main value of an affordable transport infrastructure is to guarantee a certain price limit with cost optimization, in particular over the structure's expected life span. Since this approach was covered in previous chapters – e.g. related to optimizing the care devoted to the geotechnical structures connected with different risks and applying the BIM process, which can result in cost savings – this chapter will focus mainly on the availability approach, although with the affordability approach always in mind.

5.1 Interaction of transport infrastructure with natural hazards

During the last three decades, great attention has been devoted to different natural disasters. For example, Kalsnes *et al.* (2010) wrote about 8866 disasters causing 2.3 million fatalities for the period from 1975 until 2008, during which many different and destructive natural hazards occurred. Some just for the year 2018 can be mentioned: an earthquake in Indonesia connected with a tsunami, sand storms in India, extreme droughts in South Africa, flash floods in southern and western Europe, hurricane Florence in the USA and typhoon Jebi in Japan.

Nevertheless, when the trend of fatalities due to natural hazards is studied over the last 100 years, it appears that the increase in the known number of deaths is due to the increase in the exposed population and/or the severity of natural hazards.

From the view of protection against natural hazards, it is obvious that some proactive measures are needed (Vaníček, 2011). They can be divided into three basic areas:

- Better forecast of individual event;
- Strengthening of the resistance of structures;
- Methodology for how to behave during such events.

The application of these preventive measures can significantly decrease the negative impact, as is obvious when the impact of roughly the same event (such as an earthquake) is compared for highly developed countries (as e.g. Japan) and less developed countries (such as Haiti). The significance of the aforementioned three proactive measures can be demonstrated for floods. The first point is related to floods forecast. When forecast in time and covering different possible flood-prone areas, people are better prepared in advance. The second proactive measure is associated, for example, with dike reinforcement, the construction of additional reservoirs and all measures which can increase land retention capacity. The significance of the third group of proactive measures can be shown by comparing the floods in the Czech Republic in 1997 and 2002. Although in 2002 floods affected a larger part of the country and were associated with more extreme rainfalls, 17 people died compared to 60 in 1997. This is partly attributed to better organization and equipment, especially from the view of new demands on the "Integrated Emergency System."

Nowadays, a very sensitive question is associated with expected climatic changes. The discussion of whether these changes are caused mostly by natural changes or human activities and can be significantly reduced by modification of these human activities is out of the scope of this publication. Nevertheless, it is necessary to accept the fact that some changes have been observed during recent decades and we have to react to them. A typical example is temperature increase. For the Czech Republic, we have a relatively long record of measurement. Up to the first half of the 20th century, the average air temperature was relatively stable, around 9.5 °C, but recently a steady increase of up to 11–11.5 °C has been registered, Figure 5.1.

Up to now, the amount of total precipitation has remained roughly the same, but its character has become more unsteady, with shorter periods of heavy rain and longer periods of drought.

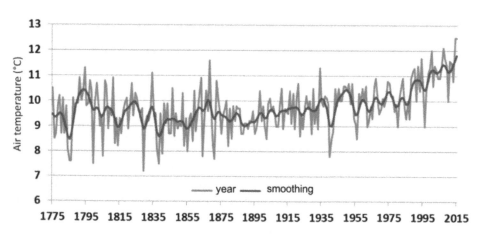

Figure 5.1 Average temperatures measured from 1775 in Prague
Source: https://urbanadapt.cz/cs/klimaticka-zmena (public domain).

This unsteady precipitation can cause more problems, either with heavier floods or with higher evaporation during the dry season. The expectation of total precipitation is different – some countries are expecting a decrease and others an increase. In any case, higher temperatures will increase sea water level, due to glacier melting, with direct impact on seashores.

Our focus from this point on is only on the interaction of transport infrastructure with natural hazards. And of the various natural hazards that occur worldwide, we will concentrate on the hazards which are typical for most of the European countries –landslides, floods and rock falls, with a few notes on other hazards.

5.1.1 Landslides

Landslides cause significant damages to both property and human lives. They create roughly 17% of all negative impacts caused by natural hazards and therefore deserve great attention.

Slope instability is typically created by:

*   Natural impacts;
*   Man-made impacts.

Concerning the natural impacts, slope instability is firstly attributed to groundwater table change (increase). In some cases, slope instability is induced by ground degradation (due to surface layer weathering) and by a change in the chemical composition of pore water. The latter is well known for quick clays, such as those in in Canada, Figure 5.2, and

Figure 5.2 Failure of transport infrastructure caused by landslide in quick clays: St Jude near Montreal

Source: www.montrealgazette.com/news/canada/Gallery+Sinkhole+swallows+home/3013315/story.html.

Figure 5.2 (Continued)

Norway (L'heureux, 2019). These problems were covered during the International Workshops on Landslides in Sensitive Clays (IWLSC), 2013 Quebec and 2017 Trondheim.

To the first case concerning natural impacts, we can also add erosion, particularly of the slope toe by a water stream or by sea waves, respective impact of other natural hazards such as earthquakes.

Cases of natural slope instability are usually very well documented by national geological institutes. Different maps of regions, countries and continents are now available to show where individual landslides are located. In many cases, individual landslides are well documented with many details. Such maps directly indicate landslide-prone areas. When a supplementary database at one's disposal, information about the type of slope instability can be obtained as well. For the case of the new transport infrastructure proposal they are creating a first source of information in such areas. The Czech Geological Survey presents documentation of ground instabilities on their web pages, see Figure 5.3 and e.g. Krejčí *et al.* (2019). In some countries, experts have started to prepare risk mapping for expected climate changes, e.g. Löfröth (2019) for Sweden.

Concerning the second case of man-made impacts, there are two main causes of landslide instability. The first is unloading at the slope bottom and the second is loading in the upper part. Damage of functionality of the man-made drains connected with resulting pore water pressure increase is another aspect in this direction.

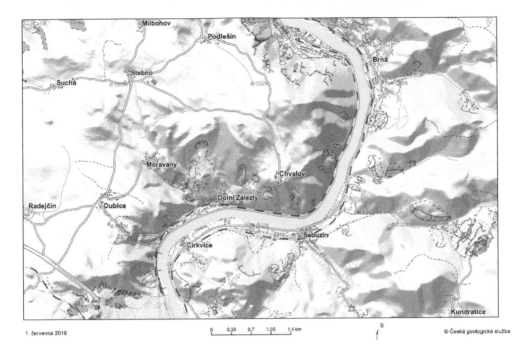

Figure 5.3 Example of landslide database

Source: Landslide map, 1:25,000. In: Geovědní mapy 1:50,000 [online]. Prague: Czech Geological Survey [cit. 2019–07–01]. Available from: https://mapy.geology.cz/svahove_nestability/ (public domain).

5.1.1.1 Transport infrastructure located within areas prone to landslides

Landslide-prone areas are marked as areas where both old and recent slope movements have been identified. The problem is that in many countries such areas occupy a large territory. For such areas, civil infrastructure engineers have to accept that reality and find the most optimal solution, which emerges from the following two steps:

- Selection of the best transport infrastructure (TI) areal placement. Knowledge of the old (recent) landslide position together with the shape of slip surface (e.g. shallow planar or deep circular) is a starting condition (Kopecký *et al.*, 2019). The manner of the landslide crossing – either as fill or cut – plays a no less important role. Ground investigation starts from the desk study (study of all existing information) and is then supplemented by *in situ* geotechnical investigation. Figure 5.4 shows proper and improper examples of fill or cut for deep circular landslides without technical support. Similarly, Figure 5.5, for shallow planar landslide, shows that fill can improve stability along the old slip surface as the ratio of the pore pressure to the total pressure decreases. At the same time, however, local stability should be verified with respect to the subsoil bearing capacity. An opposite situation is for the cut, as the aforementioned ratio is decreasing and local slope stability can cause much more problems, above all for the slope of cut in the direction against the terrain inclination.

- Selection of the best technical solution, guaranteeing stability of the whole area of interest. Starting information is the residual angle of the internal friction φ_r', which can be obtained either from testing or from back analysis of the old slip. Knowledge of groundwater pressure is important as well. Generally it is recommended to perform a feasibility study, sensitivity to individual parameters on the stability, before the final solution is proposed. The technical solution has roughly three possibilities:

    - Reduction of slope inclination, either for fills or for cuts;
    - Reduction of pore water pressure, particularly with the help of drains;
    - Construction of a retaining element.

However, new innovative repair techniques are now in the centre of interest as well (e.g. Winter *et al.*, 2019).

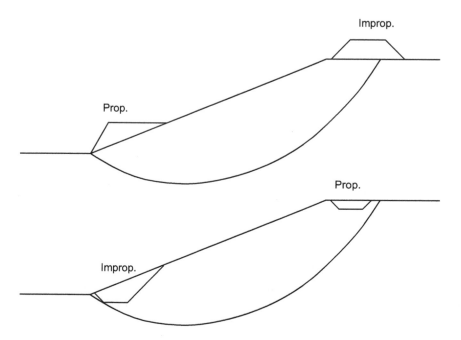

Figure 5.4 Proper and improper examples of fill or cut for deep circular old landslide

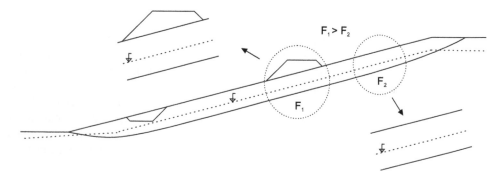

Figure 5.5 Impact of embankment or cut on stability of old shallow planar landslide

5.1.1.2 Practical case of TI interaction with old landslide

The practical example that follows stresses nearly all the aspects mentioned up to now.

Prague, the capital of the Czech Republic, is connected to other cities by eight motorways, which are arranged fanwise. Most important are those connecting the domestic motorway network to similar networks in neighbouring countries – Germany, Slovakia, Poland and Austria. Northbound motorway D8 connects Prague with Dresden in Germany. D8 is part of the pan-European transport corridor and European road E55. The motorway crosses two mountainous areas. The most critical part is roughly 13 km long, crossing the České středohoří mountains, which were declared a protected landscape area, Figure 5.6. The final version of the motorway design exploits a combination of tunnels, bridges and surface arrangements, with classical embankments and cuts situated on an inclined surface and perpendicular to the motorway alignment.

Geological conditions are very complicated there (Vaníček et al., 2018). The subsoil is composed of a complex of Mesozoic sediments such as sandstone, calcareous clay and marlstone, with small islands of older crystalline rock as gneiss. In many places, these rocks are penetrated and overlaid by tertiary neo-vulcanite rock, such as basalt. Upheaval of the older Mesozoic sediments resulted in a slope with an average inclination of about 8° downhill to the river Labe/Elbe. Quaternary materials are characterized in the central part of the slope as colluvium sediments, fine soils with high plasticity or more permeable gravels with fine particles. The total thickness of quaternary deposits in the central part is about 5–6 metres.

Two aspects were taken into consideration for the final position of the motorway:

- Area is prone to landslides, particularly to the shallow ones, with slip surface on a contact of old sediments and quaternary materials. Some slope instability was observed even for these gentle slopes when groundwater reached the surface (Pašek and Kudrna, 1996).
- Relatively good experiences with old railway lines, also passing through this sensitive region.

Figure 5.6 View on the character of the České středohoří mountains

The motorway in the discussed section was proposed in stepped cuts, in two different levels, one for each direction. Construction started in 2009 and the large landslide occurred at the beginning of June 2013, when only final pavement was missing. The landslide was described as a planar, shallow one. The length was about 500 m, so it reached an old railway track situated about 250 m above the motorway as well as a steep slope created by a spoil heap from a basalt quarry overlaying the basalt outcrop. The railway embankment moved downhill by about 50–60 metres, so that the landslide mass overlaid the motorway cuts, Figure 5.7. From this figure, it is obvious that part of the landslide below the old railway embankment moved in-plane. Trees are holding a vertical position and no cracks are visible there, while above the railway embankment the vegetation has a turbulent arrangement. The width of the landslide was about 200 m with a depth of 5–7 m, so the total volume of the landslide was about 500,000 m³. The depth of the landslide in its longest central part corresponded with the interface between the weathered surface of calcareous clay and quaternary soil cover.

The authors were involved in this problem as consultants of the investor – Czech Directorate of roads and motorway since the spring of 2014. The main aim of our involvement was to verify and observe remediation measures, to guarantee their long-term functionality for safe operation of the D8 motorway.

To be able to design the right remediation measures, it was necessary to understand why the landslide was so long. Firstly, marks of instability, just about one or two days before the landslide occurred, were observed as in the cuts for motorway, so in the upper part of the

Figure 5.7 General view on landslide overlaying motorway line

Source: Photo by Karel Pech), www.litomericky.denik.cz/.

Figure 5.8 Cracks observed on the quarry platform just above spoil heap

Source: Photo by B. Svoboda.

spoil heap, where the quarry facility was located, Figure 5.8 (Svoboda, 2014). Also, small movement of railway track was observed, leading to the speed limitation and, one day before the landslide, to closure of the railway track. These observations initiated safety precautions, so ultimately only material damages occurred.

Two questions arise from the results of observations just before the landslide occurred and subsequently from the character of the landslide:

- Can a small deformation at the motorway cut in the order of centimetres induce the deformation of the same order in the upper part of the slope – at the quarry platform? The reply is no. The answer would be yes only under the assumption that the longest part of the landslide would be rigid body.
- Could a landslide initiated at the bottom in the place of the motorway cut have caused such a long landslide? The reply is yes, but only in the case that this central part is

also close to the overall instability. In the opposite case, the landslide would end very close to the cut. A similar question and similar reply are valid for the upper part of the landslide. When the central part would be stable, the landslide initiated in the upper part – in spoil heap – would end very close to the toe of the steep slope of the spoil heap.

The main conclusion at this point was:

- All three main parts of the slope were in a state of instability at the moment the landslide occurred;
- These three parts should be treated independently, from the stability point of view as well as from the view of remedial measures;
- After remedial measures are applied to all three parts, the overall slope can be stable for long time.

CENTRAL PART

As the central part was not touched by human activity for the last century, instability was mainly attributed to climatic conditions, to the increase of the groundwater table. Hydrological data registered a long period of dry weather before the year 2009. It was period of ground investigation and motorway design. However, after that a period with higher precipitation than the long-term average was registered. Just a few days before the landslide occurred, heavy rainfalls in the wider region were registered, causing floods along the river Labe/Elbe, particularly in the city Ústí nad Labem.

The impact of the groundwater table in the upper aquifer – in quaternary deposits – is shown in a schematic cross section in Figure 5.9. Stability significantly depends on the ratio m, expressing groundwater elevation to total high. Slope stability decrease is

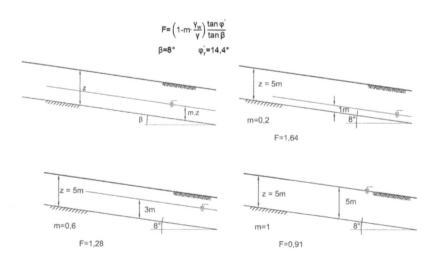

Figure 5.9 Impact of groundwater table on slope stability

presented for the residual angle of friction $\varphi_r' = 14.4°$, the only value known at that time. Later on, this value was confirmed, as the results of additional tests were in the range of 13–15 degrees. By this, it was approved that when the groundwater table is as high as the surface, stability is close to 1, reaching a point of instability. The higher water table in the landslide area is attributed not only to the extreme rainfall, but also to the collection of water from neighbouring areas (partly from the quarry platform, partly from the sides, as longitudinal landslide axis was lower than surrounding terrain – minor depression). A small note should be added about the railway embankment. As the ratio m is smaller along the railway embankment axis than for surrounding terrain, the stability along the planar slip surface is still sufficient there.

The recommended protection measures included limiting the inflow from the surrounding area, particularly from the upper part, and applying horizontal drains; these drains should guarantee that the groundwater table will stay well below the surface.

UPPER PART

The upper part of the landslide touched the spoil heap. Typical material of the spoil heap is a finest fraction obtained from the quarry, which was not very marketable. Typical fraction is 0–16 mm. The original slope was around 30 m high with an inclination close to 30°, which corresponds to the angle for free-fall filling. However, this slope was stable, as after a partial water infiltration the material had higher shear strength due to negative pore pressure. This fact allowed some of the quarry facility, even a pile of material from the quarry, to be located on the spoil heap surface (Figure 5.10).

Figure 5.10 Bird's eye view of the upper part of the slope after landslide

Source: Photo by Karel Pech, www.litomericky.denik.cz/.

For a long season of higher than average precipitation, the degree of saturation increases with a decrease of capillary forces. According to Friedli *et al.* (2017), this can cause higher pressure on the top of the central part, as well as on the railway embankment. Actually, this railway needed more frequent maintenance (particularly the track), which showed extreme deformation just one or two days before the landslide occurred.

To see the suction effect more closely, samples of partly saturated material from this part of the spoil heap were prepared in the laboratory and loaded, Figure 5.11a. However, even unloaded sample failed when saturated from the bottom, Figure 5.11b.

a)

b)

Figure 5.11 Impact of pore pressure suction on the strength of partly saturated material: (a) partly saturated; (b) fully saturated at the bottom

LOWER PART

As was already shown, any cuttings perpendicular to the slope inclination, particularly with a potential shallow planar slip surface, deserve a great attention. Roughly in the year 2009, when cuts were realized, the groundwater table in the upper aquifer was at its minimum. So the middle part of the slope was stable. At the same time due to unloading, the pore pressure dropped to a minimum, as was already described in Chapter 3 (Figure 3.23). This virtual stability was violated only locally, on places with higher soil permeability. Mostly local shallow failures were corrected, and the slope of the cut was covered by coarse material with a filter to drain water and to protect potential problems with the limit state of HYD type.

The situation was worsening with time due to pore pressure increase, either in the middle part or in the close vicinity of the cut, so the state of limit equilibrium was reached (as can be assumed from the observation of cracks) just before the landslide occurred.

To conclude, a large landslide was caused by a combination of natural and human impacts. A long wet season ending with heavy rainfall is typical of a natural impact. Unloading in the lower part of the slope and loading in the upper one are typical human impacts. About the ratio of these individual impacts, there is still discussion between many experts.

LANDSLIDE REMEDIATION

The first step taken to remediate the landslide concentrated on vegetation removal and surface levelling to decrease rainwater infiltration.

The main remediation activities started in the upper part, where the slope surface was cleaned up to less weathered rock, Figure 5.12. A drainage ditch was also installed to limit

Figure 5.12 Remediation of the upper part of slope

surface water infiltration of the central part of the slope. Later on, a deeper drainage system was executed in the central part.

A retaining system constructed just above the motorway was the most important step, allowing excavation of the sliding mass covering the motorway lanes. From the different alternatives for the retaining structure, the one composed of 13 columns made of diaphragm wall elements was finally constructed, Figure 5.13.

For the retaining structure design, the numerical calculation model was created and six design situations were defined, Figure 5.14:

- DS 1: Initial state after the landslide
- DS 2: Fill planishing + static element + anchors
- DS 3: Initial pre-stressing of anchors
- DS 4: Excavation of slipped material covering motorway up to slip surface
- DS 5: Backfill between motorway and retaining element
- DS 6: Loading berm above retaining element

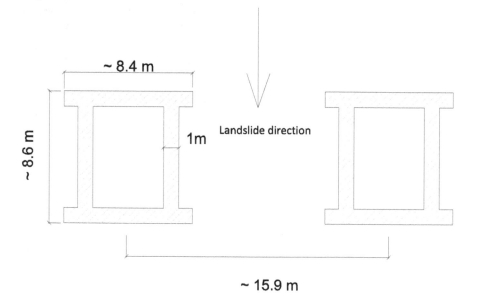

Figure 5.13 Retaining elements in the lower part of slope

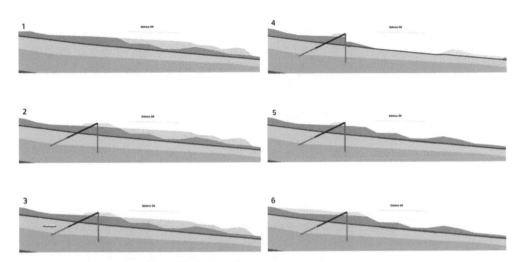

Figure 5.14 Scheme of the basic design situations

Figure 5.15 View on the opened motorway D8 together with the final arrangement of retaining elements

It is obvious that DS 4 is the most critical, as the bending moment on the retaining element is the highest. As this DS is valid only for a short time, the design was checked only for limited height of the groundwater table.

DS 6 limits any potential future slip surface above retaining element. It specifies the final state, which should be stable in the long term. After that, motorway D8 was finished and opened in December 2016 (Figure 5.15).

5.1.2   Floods

Floods are responsible for about 1/3 of damage caused by natural hazards. Range of floods and amount of damages is increasing with time, mainly due to higher usability of the flood-prone areas. Also there is duly justified discussion about climate changes and the impact of these changes on floods. With weather warming some countries are counting with roughly the same precipitation but more uneven than in history, some others count even with rainfall increase. Therefore an increased attention is fully rightful.

5.1.2.1   Flood forecasts and protective measures against floods

Flood forecasts strongly depend on the flood type. It is most difficult to forecast flash floods, which occur in a relatively small area and are caused by local extreme rainfalls (usually of short duration). The impact of this type of flood is relatively small and local. In areas where such flash floods occurred already, countermeasures can utilize these previous experiences. However, for areas without such experience, it is not so easy to define an optimal approach.

Relatively good forecasts are typical for floods in foothills, not only from the location point of view but also for specific times, as such floods are typical for the beginning of spring. The negative impact is the summation of more factors: firstly the amount of snow and ice in the mountains at the end of winter and secondly the combination of higher temperatures with rainfall. The result is a huge outflow from mountains due to melting snow and ice and rain. And the last factor is a limited infiltration of surface water to the still frozen ground. Usually forecasters have much previous experience with these types of floods, and some protection measures have already been applied in principle. They usually consist of dams, mostly in a cascade arrangement. The construction of such dams started at the end of 19th century as gravity masonry dams, and later on as fill dams. Usually they are not only providing flood protection, but they are also an important source of drinking water for cities situated in the foothills.

The last and most important type of flood are so-called regional floods, which affect one or more river basins. As they flood large areas, they are potentially the most dangerous for transport infrastructure. Also, the period of flooding is longer than for the previous two types. Therefore, material damages are extreme. Occurrence of the regional floods is associated with summer, when it is typical to experience heavy and long-term rainfalls. Weather forecasting is improving with time, so the first warning about expected heavy rainfalls can be made at least few days in advance. But this amount of time only allows for mostly administrative measures to be taken; nevertheless, the preparedness of the Integrated Emergency System is very important. Technical countermeasures at this time consist of the construction of small barriers from sand bags, and mobile barriers in large cities, Figure 5.16.

Long-term countermeasures were already started in the Middle Ages and were focused on embankments/dikes along watercourses. The height of dikes was gradually increased, and now the dikes protect against a so-called 100 years flood. This flood (Q_{100}), with an expected occurrence once in 100 years, is determined from historical records, e.g. for floods in Prague, see Figure 5.17.

Such historical records can be an answer to the question of whether the frequency of floods is increasing with time. This question is sensitive also for the Czech Republic, as

Figure 5.16 Application of mobile barrier in Prague during floods in 2002

Source: Photo courtesy of Hydroprojekt Praha.

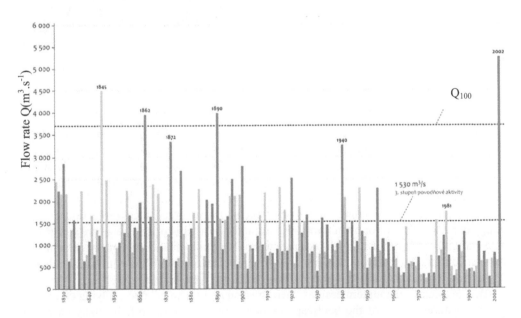

Figure 5.17 Floods in Prague: historical floods on the Vltava River in Prague during the period 1825–2002. Two piers of Charles Bridge failed in 1890 (Department archive).

Figure 5.17 (Continued)

different floods occurred there during last two decades (1997, 2002, 2006 and 2013). From Figure 5.17 it is evident that the frequency during last decade of the 19th century was also very high. As material damages and the range of information in various media are higher now than earlier, consciousness and concerns about floods among residents have increasing response.

Beyond the dikes, large and small dams make up the main technical protection systems. Large dams are typical for main rivers, while small (mostly historical) dams are constructed on small rivers and brooks. In limited cases where morphological conditions and manners of land utilization allow, dry polders are used. Theirs reservoirs are filled with water only during floods, just to discharge this water volume later on to the main stream.

Besides technical countermeasures, there are also non-technical ones, relying on increasing the natural ability of the landscape to slow down water run-off. In many cases, the discussion about the preference of these two basic approaches is quite vivid, but many arguments were put forward that prefer the technical approach (Disse, 2015). However, on the non-technical side, such as agricultural steps, there are some additional benefits, e.g. limitation of landscape erosion.

5.1.2.2   Impact of floods on transport infrastructure

In principle, there are two basic impacts of floods on transport infrastructure:

- Interruption of transport infrastructure utilization due to its flooding; after flood level downturn, this TI can be used again.
- Disturbance or complete failure of transport infrastructure, requiring partial or complete reconstruction; resulting in shutdown for the period needed for this reconstruction.

The first case is sensitive for bridges, where reopening is connected with a structure check, not only for the upper structure but also for the pier's foundation, as higher seepage velocity can cause scouring up to uncovering the footing bottom, e.g. FHWA (2001). Such a scenario is sensitive especially for historical bridges, where the foundation depth was relatively shallow. A typical example is Charles Bridge in Prague, constructed in the 14th century. With time, individual foundation piers were destroyed during heavy floods, for the last time in 1890. Figure 5.18 shows a historical bridge from the 13th century in the town of

Figure 5.18 Historical bridge in Písek during and after the floods in 2002

Source: https://cs.m.wikipedia.org/wiki/Soubor:Pisek_povoden.jpg (public domain).

Písek on the Otava River, which was completely flooded in 2002, but survived practically without any damages. A detailed description of the interaction of bridges with floods on the Po River in Italy is presented by Malerba (2011).

Another exceptional case should be mentioned. Due to overflowing of large dams situated on the Vltava River combined with an extreme flow rate from the Berounka River, the flood wave hit the capital city of Prague. The town district of Karlin was completely flooded, with negative impact not only on road and tram transport but mainly on the metro system, Figure 5.19. Many stations were flooded, causing the interruption of transport for several months, as controlled pumping up of the water followed by lining control was a very complicated process (e.g. Romancov, 2003; Soga *et al.*, 2011).

Figure 5.19 Floods in Prague in 2002, affecting metro, bus and tram transport systems
Source: Romancov, 2003.

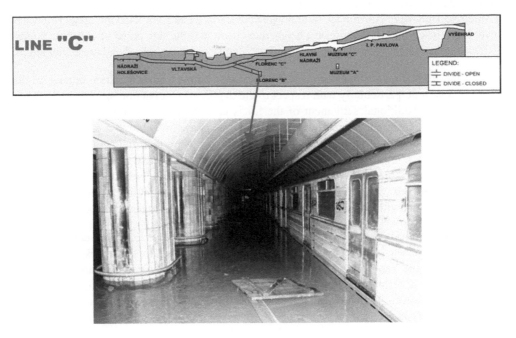

Figure 5.19 (Continued)

The second case, disturbance or complete failure of transport infrastructure, deserves more attention. Briefly speaking, the following are possible interactions between floods and transport infrastructure, which can result in disturbance or complete failure:

- Disturbance of small (historical) dams, whose crest is used for TI or whose failure can initiate problems on the TI;
- Disturbance of dikes;
- Disturbance of TI crossing waterway;
- Disturbance of TI located on a contour line when surface water flows down perpendicularly to it;
- Disturbance of TI following watercourses.

Herein it is useful to mention another specific case – interaction of floods with TI in a phase of construction. During the construction of TI in flood-prone areas, it is necessary to take the possibility of flooding into account. Principles of the risk management process are similar as for the construction of hydro structures protected by cofferdam.

5.1.2.3 Interaction of small historical dams with transport infrastructure

About 75,000 small dams were constructed in the Middle Ages, roughly between the 14th and the 16th century in the area of today's Czech Republic. The purpose of such dams was multifunctional – mostly for flood protection, but they also had a positive impact on the environment (particularly for moorland). Reservoirs were used for fisheries, sometimes even as energy source for water mills. Due to different demands for the land utilization, as well as the failure of some dams, only about 25,000 still exist. The height of these dams is roughly between 2 and 12 metres. They were constructed as homogeneous earth dams from local material. The water level in the reservoir is hold now on a height that guarantees the safe outlet of 100-year floods.

During the last few decades, a number of problems have been raised and are attributed to:

- Higher frequency of transport over the dam crest with a higher dynamic impact on the embankment, particularly with respect to the degree of compaction applied in the Middle Ages. Additional settlement of the fill is causing visible non-uniform settlement of the dam crest. But more potentially dangerous is the invisible loss of the contact of the old wooden bottom outlet with surrounding soil. A preferential path for leachate is established, and subsequent internal erosion could lead to total dam failure.
- Substitution of the old wooden underwater discharge system with new concrete gullet. Nappe is mixed with air causing cavitation in the wooden outlet, with a negative impact on the contact with soil or on the connection between individual segments of this outlet.

Most critical from the complete failure point of view, however, is the so-called domino effect of failure caused by the cascade arrangement of individual dams (Vaníček and Vaníček, 2004; Vaníček and Vaníček, 2008). With respect to the dam density per unit area, this is not atypical for the Czech Republic. The authors obtained practical experience of the cascade arrangement of seven small dams on the small river Lomnice in South Bohemia, Figure 5.20. When the first dam, particularly with the highest reservoir volume, failed during floods with

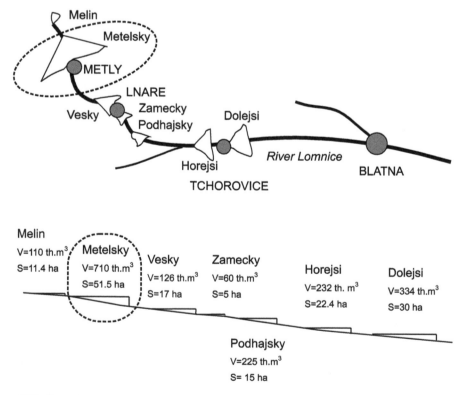

Figure 5.20 Cascade arrangement of small dams for the small river Lomnice, leading to the domino effect of failure

Figure 5.21 Damage of the dam with bituminous carpet on its surface caused by surface erosion

a flow rate Q close or even higher than Q_{100}, the other dams, situated below, have a limited chance to survive with respect to the additional flood wave from the failed upper dam.

From the experience obtained during the investigation of failures as well during the dam's reconstruction, the following points can be underlined:

i. Historical files revealed that a similar domino effect of failure occurred there during floods in 1895. All seven dams failed at that time, and the recorded level of floods was a little bit lower than in 2002. In 2002, five dams failed completely but two were only strongly damaged. Both that survived had bituminous carpet on the crest, as important roads pass over these crests, such as road E49, Figure 5.21. So this means that the crest reinforcement can significantly slow down surface erosion, even when the maximum flood level was about 2 metres over the crest.

ii. Observation and back analysis showed that the ultimate limit state of the HYD type was the main reason for failure, particularly caused by surface erosion. But for two cases internal erosion was responsible as well. However, a human factor played an important role in these two cases. An old wooden outlet was left in the dam body during reconstruction at the end of 18th century. In the second case, the dam was partly "improved" by sealing wall made by jet grouting only two decades ago. The preferential seepage started along the bad contact of the thin sealing wall with the bedrock composed of great boulders. Increase of hydraulic gradient at this place was important as well.

iii. During the dam's reconstruction, two questions were rather interesting. First, can we use the same material as was used by our predecessors when index properties (grain-size distribution curve and low plasticity) reveal that, according to the standards for small dams, the soil (granite eluvium with grain-size distribution curve similar for sand with limited amount of fine particles) is not appropriate for a homogeneous dam ? The final reply was yes, as permeability tests showed a relatively low filtration coefficient, in the order of 1×10^{-8} to 1×10^{-9} m/s, probably as the result of a high content of mica typical for granite eluvium. The second question was focused on the lowering future risk caused by surface erosion. Additional unprotected but reinforced spillway was proposed with a crest lower than the remaining crest of the main dam. This means that the overflowing will start at a place selected as most appropriate and, because of reinforcement, will cause less damage.

iv. Dams and similarly transport infrastructure crossing waterways are usually designed for flow rate corresponding to Q_{100}. However, the additional wave from a breached dam can significantly increase this flow rate and can create potential threat for a dam or transport infrastructure situated below. The classification of such a dam into a higher consequence class category is obvious, as after that its operation needs a better checking system. To be able to recognize the potential risk of these dams, we started with numerical modelling of the outflow from the breached dam and subsequently modelled the transformation of this additional flood wave as a function of time and path, to figure out the strength (peak flow rate) of this additional wave when it reaches crossing transport infrastructure or another dam within the cascade arrangement (Vaníček *et al.*, 2016).

5.1.2.4   *Interaction of dikes with transport infrastructure*

This type of interaction is most sensitive for the majority of countries. A great deal of historical data points to significant damage caused by dike failures. For example, the failure of dikes along the Danube River below Bratislava in 1965 destroyed 3910 houses, and another 6180 were damaged. Data for the Morava River during floods in 1987 are more specific to transport infrastructure – 946 km of railway track and 1850 km of roads were strongly affected. The 2D effect plays a very important role, as dike breaching will start at the weakest point. A short length of failed dike can completely cause the malfunction of a long dike protection system, Figure 5.22.

Figure 5.22 Danube River below Bratislava in 1965 – relatively small breach caused large area flooding
Source: Department archive.

These types of failure are similar to those of small dams, as they also overflow, and surface erosion predominates. Therefore, countermeasures are also similar, for example installing bituminous carpet on the dike crest and reinforcing selected places where overflowing can start. One method of reinforcement was tested in the laboratory (Vodička *et al.*, 2008). Between compacted layers of soil were implemented layers composed of a mixture of construction and demolition waste, cement, and geosynthetics short fibres, also classically compacted. Such a dike survives even long-term overflowing with only minor damage, Figure 5.23.

Other types of failure depend on the composition of the subsoil. Failure by piping is typical for the permeable layer, while slope stability can occur for very soft ground. Additional berms not only improve slope stability but also by elongating the seepage path they decrease the hydraulic gradient. However, when the first marks of piping are visible during floods, very often sand bags are applied around the places with water emergence, to decrease the hydraulic gradient and to slow down particle erosion. After such problems a complex solution is often proposed, consisting of the design and execution of a vertical sealing element such as a sheet-pile wall or a thin cement-bentonite sealing wall.

Figure 5.23 Reinforced dike with higher resistance against surface erosion

Source: Photo by J.Vodička.

5.1.2.5   Other interactions

The examples of TI interaction with waterways or with surface water have many similarities. When the capacity of a culvert aqueduct is not sufficient (very often it is blocked by rubbish), the water level rises and TI embankment is loaded similarly as a small dam or dike. The failure can occur in the part of embankment close to the culvert, Figure 5.24.

The authors initiated discussion in 2018 concerning the type of failure (Vaníček *et al.*, 2018). When the soil is very permeable, the seepage is rather quick and the outflow on the downstream face can initiate internal erosion. When the soil has low permeability, seeping water cannot reach the downstream face, as the time of flooding is relatively short. Nevertheless, after the quick flood wave decrease, the water inside of the embankment can create

Figure 5.24 Photos of failed railway embankment: (a) the interaction with the little river Lomnice (Vaníček, 2004); (b) the interaction with surface water, Bari–Taranto railway track (Burdo, 2016)

slope instability of the upstream face by reverse outflow. As often the character of failure did not correspond to either of the two aforementioned cases, attention was focused on soil with average permeability, roughly of the sand type, which was originally unsaturated. As floods are induced by heavy rainfall, water infiltrates into the embankment body from all sides simultaneously and air bubbles are trapped between these saturated zones. The air pore pressure gradually increases and is finally capable of lifting the downstream face. The displacement (slide) of this face can evoke internal erosion leading to total collapse. The ability of pore air pressure to lift up a saturated part was simulated in the lab, Figure 5.25. Dry sand was covered by a saturated layer, and via capillary forces saturation also started from the bottom. Increasing pore air pressure between both of these saturated parts was finally able to lift up the upper layer. By this simple experiment, the theoretical scenario for embankment failure was proved.

The recommendation in this direction leads to the application of a soil type that is appropriate for homogeneous dams – one with low permeability and good shear strength.

The last case of TI interaction with floods is typical for narrower valleys where TI follows a water stream. During floods, the water level is rising, as is velocity. The sides of the adjacent TI are subject to erosion by higher energy leading to the stoppage of the transport there,

Figure 5.25 Upheaval of saturated sand layer by pore air pressure increase in partly saturated sand

Figure 5.26 Typical case of damage of a road following a water stream
Source: Photo by HZS Zlínského kraje.

Figure 5.26. The countermeasures are more complex. They have to start in advance, and the evaluation of places with the highest expected erosion energy is not a difficult task. Reinforcement can guarantee higher resistivity when floods will start. When failure already occurred, there is strong pressure on quick reconstruction from residents as well as politicians. The problem is that under this pressure, standard (traditional) methods of reconstruction methods are applied, usually with the same resistivity. Therefore, new methods incorporating the principles of affordability and availability (generally more robust) should be prepared in advance to reduce the probability of failure at the same place in the future, during the next floods.

5.1.3   Rock falls

Rock falls create another important threat to the transport infrastructure. Risk resulting from this natural hazard is increased by its suddenness; usually it is an unexpected event.

The only "advantage" is that the place where such an event can occur can be predicted with a certain probability. Such places are typical for:

- Narrow valley with a watercourse where transport infrastructures (motorways, railways) are situated between the river and rock slope. Such a case does not require high elevation above sea level, Figure 5.27.
- Narrow neck between rock mass allowing the only access to the mountain valley.
- Transport infrastructure passing over mountains, and typically when crossing a high and extensive mountain range.

Figure 5.27 Typical case of narrow valley with watercourse: Mariánská rock in Ustí nad Labem – Czech Republic

Source: Photo by Václav Hribal, www.turistika.cz/mista/marianska-skala/foto?id=425107.

The character of rock movement is different when compared to soil, from rock sliding to over-toppling to free fall from a cliff. The trajectory of falling rock is different, not only straightforward but also close to jumping. Falling rocks are the result of morphology, geological profile and climatic conditions. The negative impact of geological profile is presented in Figure 5.28, where it is obvious that layer orientation and number of discontinuities contribute to the risk of rock falls (Záruba and Mencl, 1969). The increase in the frequency of rock falls during heavy rainfalls or during early spring thawing points to climatic conditions. The discussion to the impact of vegetation is still open, as in some cases this impact is positive – vegetation carpet can decrease the fall of smaller rock particles, or trees roots can protect rock falling. On the other hand, roots have a negative impact by opening cracks, and after tree fall the potential risk is higher.

As the character of individual regions with a high potential risk for rock falls is different, the experiences are also different. The proposed countermeasures therefore should proceed from local experiences but also from the fact that transport utilization is increasing, so that the probability of direct threat from falling released rock is and will be higher. Therefore, the proposed countermeasures should be improved with time, with the help of new technologies.

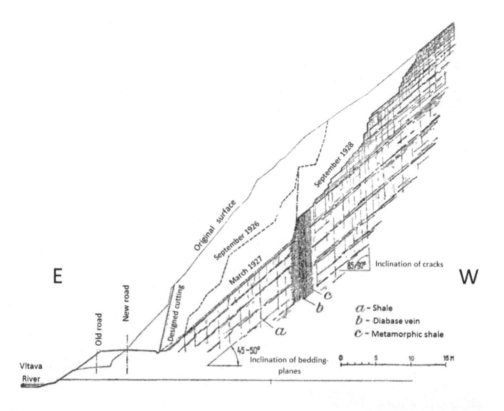

Figure 5.28 Negative layer orientation and number of discontinuities as the result of rock falls

Source: Department archive.

The individual steps for safe and economic design can be divided into:

- Evaluation of existing experiences – not only from the technical perspective but also from the possibility to define risk in accordance with Equation (2.2);
- Investigation performed for a given rock slope;
- Protection measures design, whereas basic differentiation is between protection measures applied directly on rock slope, respective protection measures applied at the slope toe, just in front of transport infrastructure;
- Proposal for monitoring system and for the maintenance of protection measures.

These steps will be specified for one practical problem, namely for the protection of a road between Prague and the small city of Štěchovice (Jirásko and Vaníček, 2015, 2017; Vaníček *et al.*, 2018).

5.1.3.1 Practical example – individual steps for safe and economic design

The road between the capital of Prague and the smaller city Štěchovice passes along the Vltava River, roughly in the south direction. The most critical place is situated close to a concrete dam with hydraulic power station. Due to a high water level in the reservoir, the possibility for road positioning – which demanded slope toe undercutting – was highly reduced (see Figure 5.28 taken during road construction in 1928).

Evaluation of existing experiences – After about 90 years of using the road, there are many experiences. Foremost is the geological profile with unfavourable orientation of individual layers – see Figure 5.29.

Figure 5.29 View on the rock slope with an unfavourable layer arrangement and the road situated close to the water reservoir

Each year several accidents happened there. Protection measures were focused mostly on a small retaining wall combined with wire netting, situated at the slope base. It was not sufficient for larger boulders or for smaller ones which jumped over this protection system, Figure 5.30.

As the number of interactions of vehicles with falling rocks increased (mostly due to higher transport frequency), the protection measures needed to be improved.

Rock slope investigation – This is very complicated, particularly from the access point of view. Visual observation is needed, sometimes even with the help of climbing technique. The main aim of this investigation is to determine the places with a high potential risk of falling. Two basic cases were denoted as most critical – Figure 5.31:

- Large blocks, with high potential for overturning and sliding;
- Individual layers which were not supported from the bottom – sliding along existing inclined layer is in this case extremely critical.

Figure 5.30 Direct impact with road

Figure 5.31 Blocks with expected sliding and rotational movement

However, the investigation was not only focused on the most sensitive places, but also on what protection measures could be initially proposed. Such a proposal should also include the real possibilities at the slope bottom. Roughly, the following possibilities were discussed:

- Controlled unblocking or disintegration during closed transport on the road.
- Monitoring of critical large blocks – when signal of monitoring with predefined critical deformation value is transferred to the control centre with the possibility of reacting quickly to treat and close the road.
- Selection of parts and profiles where protection measures can be applied on the site (on the rock slope).
- Selection of profiles where protection measures are recommended just in front of the road.

The final output of the investigation is summarized for observed slope in graphical form, where individual observed points are displayed (with the inclusion of more specifications in written form supported by photos), together with a demarcation of protection measures by different colours, Figure 5.32.

PROTECTION MEASURES DESIGN

Steel mesh anchored to the surface or dynamic barriers are the main measures applied to the rock slope (Brandl and Blovsky, 2002). In the first case, the released stones (e.g. due to weathering) are held in place and weathering can go on. In the second case, falling stones are caught by a dynamic barrier. Different software programs able to model the stone trajectory and kinetic energy are now used for the design of barrier height and its strength. Such an example is shown in Figure 5.33.

Different retaining walls are preferred for measures applied at the slope toe, when there is a sufficient volume for captured stones. The similar modelling as for the previous case can be used. Vaníček et al.(2018) described, for example, a retaining wall from reinforced soil where wall surfaces are created by old tyres filled with soil able to absorb some energy of falling stones.

MONITORING AND MAINTENANCE

Two basic approaches are taken for monitoring the rock slope. Monitoring directly on the rock slope is oriented on larger blocks for which protection measures would be very problematic. Different sensors, such as extensometers and tilt meters, are applied there. However, signals from such readings should be translated to the collection point, where they are evaluated with recommended action – in the most critical case connected with closing the threatened transport infrastructure, Figure 5.34. The authors for such a case used a similar system as was used for the metro lining monitoring when the access was limited (Soga et al., 2011).

Monitoring of the slope deformation over a long distance is very attractive. From the monitoring of individual places, there is a marked transfer to the monitoring of a whole area. A similar transfer occurred for metro lining monitoring, where for long sections the whole tunnel profile is monitored. In addition to new methods of monitoring, laser scanning or computer vision can be mentioned (Soga et al., 2011). For rock slopes, ground-based radar

Figure 5.32 Final output of the investigation with recommended types of protection measures

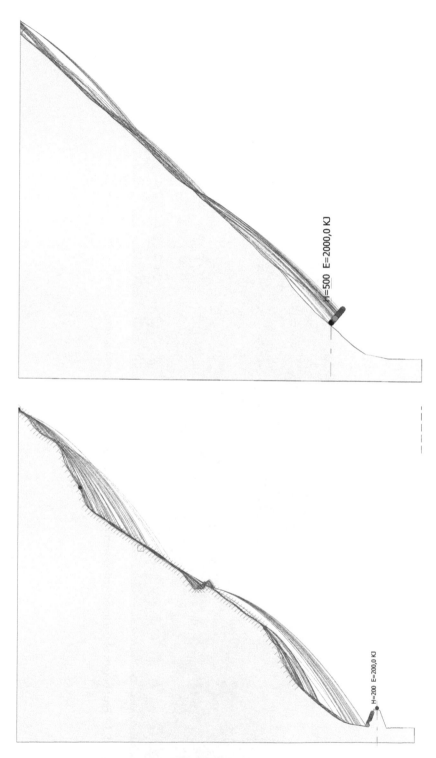

Figure 5.33 Modelling of the released stone trajectory and energy in GeoRock software

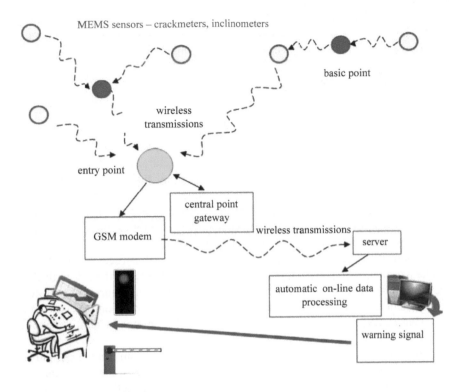

Figure 5.34 Recommended warning system for rock slopes

interferometry (GBInSAR) (Barla, 2014) can be used, which can distinguish deformation changes on the order of one-tenth mm.

Maintenance should be planned in advance and is focused on removing captured boulders and controlling the deterioration of protection measures.

5.1.4 Other natural hazards

Up to now three main natural hazards with respect to the interaction with transport infrastructure were discussed. However, there are many others, which are not typical for Europe, but for some other continents or for specific countries or regions can be very sensitive.

Typical geological hazards include:

*   Earthquakes, volcanic eruptions, avalanches, mudflows, lahars, costal erosion and sinkholes.

Meteorological and climate hazards include:

*   Hurricanes, tornadoes, cyclonic storms;
*   Blizzards, hailstorms, ice storms;
*   Drafts, heatwaves.

The combination of individual forms is typical for natural hazards, e.g. an earthquake can be the cause of landslides, rock falls and mudflows in mountainous areas. Sliding mass can block water streams, creating natural dams. Increasing water pressure from impounded water in a reservoir can destroy this natural dam causing floods in areas below it.

As these others natural hazards are typical for specific regions, the problem of their interactions with transport infrastructure is emerging from recent experiences. They demand detailed solutions, which are out of the scope of this publication. Nevertheless, short general notes are added.

Earthquakes are caused by sudden movement of tectonic plates, resulting in a sudden release of energy and thus creating seismic waves. For the design of civil engineering structures, including the earth structures for transport infrastructure, the seismic impact intensity is the deciding factor (Towhata, 2008). For specific regions, this intensity can be defined as a probability of occurrence. Maximum peak acceleration in horizontal direction (as percentage of *g*) is after that defined for expected seismic impact intensity. Slope stability calculation should account for this additional horizontal loading, see for example the slope stability method proposed by Sarma (1973, 1979), or Janbu (1973).

Observation of the impact of earthquakes on transport infrastructure, e.g. for the Kobe earthquake Koseki (2010) proved the usefulness of geosynthetic-reinforced earth structures. The probable reason consists of the significant difference between short- and long-term geosynthetics tensile strength, when the design is performed for lower long-term strength, while seismic event is representing short-term loading, when higher short-term strength can be utilised.

For significant earthquakes, the movement of tectonic plates can create large faults on the earth surface, with a negative effect on transport lines, e.g. the San Francisco earthquake in 1906. Strong submarine earthquakes can cause tsunamis, with negative impacts on seashore.

A *volcanic eruption* expels various materials. It starts with large and high clouds containing only small particles (with a negative impact on air transport) and ends with the expulsion of large boulders or high temperature lava/magma, which could potentially interact with transport infrastructure.

Avalanches, mudflows and lahars are typical for mountainous areas, when a mixture of soil/rock particles with snow or water slide down, sometimes with extreme velocity. Sliding is caused not only by gravity but also by heavy rainfalls or the melting of snow in the spring, whereas lahars is the term for such sliding caused by the eruption of a glaciated volcano. Such flows can destroy entire towns or transport infrastructure in seconds and kill thousands of people.

Typical countermeasures consist of the construction of barriers on mountainsides, most often in a cascade arrangement. Such barriers should slow down the speed of movement and bring to a stop the movement of the stiff particles. In the past, a combination of timber and stone were used, e.g. in the form of stony cribwork. Later, small gravity dams constructed of either masonry or concrete prevailed. Nowadays, permeable barriers are preferred, constructed from reinforced coarse soil or rock gabions, or using some sort of barriers typical for rock falls. An overview of the different types and the design approach taken is presented, for example, by Hofman (2019) or Takahashi (2019).

Sinkholes are potentially very dangerous as they can arise very quickly and unexpectedly. They are caused by the vertical upward propagation of underground caverns. The underground caverns can be natural, created by geological evolution, or artificial, caused by human activities. For the first case, a road along the Dead Sea in Israel can be mentioned.

The propagation of natural caverns up to the surface is the typical case there; these caverns are created by the dissolution of salt sediments by seeping water.

Mining areas are typical for the second case, especially when mining was fairly shallow and the exact situation of excavated parts was not well documented. This fact was taken into consideration during the preparation of high-speed train track close to Leipzig, Germany (Alexiew *et al.*, 2002). The contact of the embankment with subsoil was strongly reinforced to enable, for a limited time, the embankment weight to be carried with very low deformation on its surface, even when a sinkhole above the mined area was created. By using this countermeasure, the operation of the railway track doesn't need to be interrupted. Signals from sensors positioned on the reinforcing element can relay information about this specific problem and its location. After that, the sinkhole can be injected without delay.

Hurricanes and similar strong winds can force surface water waves against the seashore, increasing the storm surge level significantly above its general level. For low land, typical for The Netherlands, this situation is extremely dangerous. In 1953, the dikes along 190 km of seashore were overflowed. This resulted in catastrophic flooding, through nearly 90 breaches, with 150,000 ha of polder land inundated, causing the death of 1850 people (Pilarczyk, 1996).

In such situations, when the storm surge level increases, large river estuaries are extremely sensitive as river water cannot outflow to the sea. The water table in an area close to the estuary rises, causing flooding. Protection measures for this scenario – e.g. construction of barriers protecting the river estuary against storm surge –are typically used in The Netherlands. The sensitive discussion is connected with expected climatic changes, sea level increase and land subsidence (van Staveren, 2006; Beetstra and Stoutjesdijk, 2005).

Southern seashores in the United States are also susceptible to hurricanes, namely in the Mississippi river delta. In September 2005, hurricane Katrina sharply increased water level, causing the failure of protection dams of lake Pontchartrain and of dikes along the Mississippi river. Nearly 80% of the city of New Orleans was flooded, resulting in 1037 deaths and over 200 billion US dollars in damage.

Drought, heatwaves –these climate hazards are also potentially dangerous to transport infrastructure. Average temperature increase is changing our standards for the protection of ground against drying and the development of shrinkage cracks. Dry and hot seasons cause the damage of the green protection of slopes, opening the way for surface erosion when heavy rainfalls start.

5.2 Interaction of transport infrastructure with man-made hazards (accidents)

Not only natural hazards but also man-made ones, sometimes simply called accidents, can have a negative impact on transport infrastructure utilization, particularly from the availability and affordability point of views. In this section, not all accidents will be deal with, but only those with some relation to the discussed problem. Many of them, however, are the result of a combination of factors, similarly as for natural hazards.

A classic example is that of a traffic accident. In most cases, it is an isolated traffic problem. However, when there is the collision with a civil engineering structure – such as a bridge pier – the problem is more serious, as the bridge stability will need to be assessed by a structural engineer. Similarly, when the collision is with retaining structure that enables a steeper slope, a geotechnical engineer is responsible for such assessment. The question arises as to whether this type of collision should be part of the design, as one of the design situations. If so, the engineer needs to know the expected frequency of such an accident as a function of energy of the collision with the civil engineering structure. After that, the selected approach is very similar to the design for natural hazards, such as when dikes are designed for a hundred-year flood. Similarly, a civil-engineered structure can be designed for the accident causing additional loading which can be expected with a certain frequency.

In the next sections, three cases will be discussed in more detail. They focus on the decrease of potential negative impacts of subsoil contamination, crashes with animals, and finally fire.

5.2.1   Prevention against subsoil contamination

There are roughly two basic possibilities leading to the potential contamination of subsoil.

The first one is caused by the gradual leakage of fuel and exhaust from various vehicles – cars, trucks, trains etc. – and can be sensitive not only for motorways and railways but also for parking areas and airports. The main aim is focused on the collection of the mixture of potential contaminants with rainwater in a sealed reservoir, Figure 5.35. Subsequent cleaning will allow the discharge of water into a watercourse or its infiltration into subsoil.

For areas with a bituminous carpet or a concrete layer, the surface water will be retained by peripheral drain and subsequently carried to the reservoir by different standpipes. A sealing element on the contact of ballast or sub-ballast with subgrade is preferred for railway lines, e.g. application of geomembrane or bentonite mattresses.

The second source of potential subsoil contamination is caused by a traffic accident of the truck transporting liquid chemicals. A larger amount of liquid chemicals requires a quick solution, for which detail information about chemical composition and about composition

Figure 5.35 Retention ponds during construction and operation

Figure 5.35 (Continued)

of subsoil (ground model) are needed, e.g. particularly the groundwater level and its fluctuation and the permeability of individual layers of ground. In this respect, the BIM process is very useful, both from the position point of view and from that of ground properties. Such quick information can help to conservatively specify the depth which can be contaminated for given time, subsequently excavated and treated off place.

For a more precise approach, an estimation of individual factors influencing contaminant spreading (see Equation (4.7)) is needed. In this case, two advanced methods for the contaminant of spreading can be applied – for unsaturated ground or for seepage along preferential paths – as the subsoil is not homogeneous; it is an environment which includes different preferential paths along which seepage happens much faster (Císlerová *et al.*, 2011).

5.2.2   *Prevention against interaction with animals (eco ducts)*

Problems associated with negative impacts of crashes of vehicles with animals are getting relatively more attention. Higher speed as well as higher transport density are decreasing the possibility of eliminating potential crashes and are causing more human and material damages, together with the subsequent impact on transport line utilization. The classical approach of constructing fences along the transport lines very often interferes with the demands of environmentalists who prefer the interconnection of both parts for the free movement of various animals. Such a solution involves the construction of communication paths either in the form of bridges or underground passages.

A culvert aqueduct was mentioned already for the diversion of water, particularly for floodwater. For the passage of animals, a new term is now preferred – eco duct. To find a cheaper solution than classical concrete aqueducts, different possibilities are proposed. In principle, they incorporate more ductile structures. For small animals such as frogs, such structures incorporate polyethylene pipes, corrugate pipes and flexible culverts. For larger animals such as red deer or bear, corrugated metal conduits are preferred.

For such large eco ducts, it is necessary to pay attention not only to the shape of the corrugated metal conduit (with predominant compression loading) but also to soil compaction. Soil compaction should be symmetrical, preferably at the same time from both sides. The application of soil reinforcement alongside and above conduits can decrease loading (Vaníček and Vaníček, 2008), Figure 5.36. Detailed specification for the design of different types of pipelines or corrugated metal conduits is given in special publications, e.g. Selig (1985) and Moser (2001).

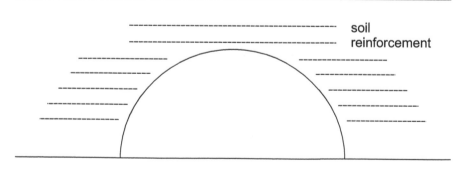

Figure 5.36 Application of soil reinforcement to eliminate asymmetrical loading of corrugated metal conduits

5.2.3 Prevention against fire

The interaction of fires with transport infrastructure is very sensitive for tunnels, as it can block transport for a relatively long period. Fire mostly occurs because of traffic accidents, but exceptionally as the result of spontaneous combustion of an individual vehicle.

For the transport infrastructure embankment or cut, there are better possibilities from the view of direct reaction to an incurred fire – usually quicker information, better accessibility for the fire brigade. Better possibilities also exist concerning the quick removal of the burned wreck from the main transport line. The direct technical negative impact is mainly attributed to the high temperature of fire, with questionable interaction with the road surface. There is also potential risk that the fire will spread to the surrounding environment, particularly when the vegetation is dry.

Fire is usually combined with the leakage of driving fuel, and therefore the following problems should be taken into account:

* Outflow of burning fuels to the drainage system, which can be after that non-functional, especially when plastic tubes are involved;
* Outflow of burning fuels on the embankment slopes, with negative impact on protection system against surface erosion, such as vegetation or different anti-erosion mattresses, plastic or natural (e.g. coconut matting);
* Outflow can also penetrate the ground, causing ground contamination – which should be treated in accordance with that offered in section 5.2.1.

The direct impact of fire on soil properties is discussed by Pereira *et al.* (2019).

Note: For reinforced slopes or reinforced retaining walls, where different geosynthetics are used, the designer usually counts with their protection by soil, at least by few centimetres. The question is if this protection can protect the geosynthetics against failure by high temperature (fire), namely in the case when the protection layer was partly washed away.

Briefly speaking, protection of earth structures against fire have to start in the design phase by considering the aforementioned possibilities. The design should guarantee that flammable parts will not have a chance to come into direct contact with fire/high temperature during all the construction and service life phases of the designed earth structure.

Chapter 6

Conclusion and final recommendations

New demands defined for transport infrastructure can be compiled into two basic aspects – incorporating engineering and sustainability. Addressing these two aspects for earth structures for transport engineering introduces the following main objectives:

• Securing structure safety for its expected lifetime and, at the same time, to meet economic demands;
• Securing an environmentally friendly approach;
• Guaranteeing transport infrastructure operation (with limited restrictions) when affected by natural hazards or during accidents caused by human factors.

Attention is therefore focused on identification of all potential risks, on theirs's evaluation with subsequent design and structure construction and maintenance, leading to the fulfilment of the main objectives. We hope that the presentation of the information in this volume is straightforward for all partners of the construction process.

In the basic form of presentation, attention is focused particularly on individuals responsible for planning the transport infrastructure, investors (owners) and environmental protection specialists. The aim is to show clearly that every civil engineering structure interacts with and affects the environment/ground. This interaction is even more sensitive for the earth structures for transport engineering, as it utilizes ground as a main structural material. Therefore, we feel it is necessary to stress that our main aim is to ensure that this interaction is environmentally friendly.

The basic interaction problems between the proposed structure and the environment/ ground and between engineering and sustainability objectives should be identified at the end of the preliminary phase of geotechnical site investigation. The results of this identification can be useful in two basic aspects. Firstly, for requirements of the earth structure design, what might the ground response to the proposed structure be? Respectively for requirements of the EIA (environmental impact assessment), what would the impact of the earth structure on surrounding ground be?

In the detailed form of presentation, attention is focused on persons involved in individual phases of the particular earth structure design, construction and operation. Potential hazards and their consequences are identified together with the appropriate responses and mitigation actions. As the ground is part of the natural environment (as well as a construction material) with heterogeneous properties, the geotechnical risks are relatively high. With respect to the very small relative proportion of the environment which can be investigated and tested, a full understanding of the geotechnical risk is impossible.

Sometimes we are speaking about the possibility of investigating only one millionths of the ground volume affected by proposed structures.

A generally valid principle (particularly for civil engineering structures) is that the range of care devoted to the collection of information, to the design and finally to the actual construction is strongly associated with risk. With respect to this principle, geotechnical structures are differentiated according to the potential risks, or according to the potential consequences of failure. The selected approach then corresponds to the level of risks and consequences.

Eurocode 7 Geotechnical Design, which is valid in Europe, recognizes three geotechnical categories (GCs). These categories are defined as a product of consequence classes and geotechnical complexity.

With respect to the design and structure performance, four main phases are distinguished and discussed:

- Ground model – the result of geotechnical/ground investigation;
- Geotechnical design model – the result of cautious selection of characteristic geotechnical parameters, representing the investigated environment/ground;
- Calculation model – ensuring the structure safety and economical effectiveness, when the principle of limit states is preferred – ultimate limit state and serviceability limit state;
- Structure execution – ensuring the structure is constructed according to the prescribed demands.

For simpler cases (for lower GCs – GC 1 and GC 2), the recommended procedures for individual phases are specified, which in principle correspond to the requirements presented in Eurocode 7 Geotechnical Design. Requirements are specified either in the first- or second-generation version, whichever is valid at the start of the design process.

For more complicated cases (GC 3), basic principles to their solution are indicated, but mostly they are supplemented by links to the literature where more information can be found.

The practical use of presented approaches is shown using real examples in which the authors were involved, namely for cases where failures occurred. The described cases indicate the range and reasons of failures, as well the steps used for reconstruction. These practical examples incorporate a geosynthetic-reinforced soil, an interaction of a motorway with a landslide and an interaction of earth structures with floods. The last case includes recommendations for increasing the safety of an unstable steep rock slope.

The four key phases are also interconnected by principles of sustainable construction and principles of availability and affordability. The principles of sustainable construction (as is decrease of land consumption, energy consumption or natural construction aggregates) should be accepted from the start. The investor (owner) should include these principles the bidding process, stating that they will be evaluated for the entire life span of the structure. The same is valid for the principles of availability and affordability. Higher structure robustness can fulfil these principles. The greater resistivity of the structure during interaction with natural hazards or accidents caused by human factors would decrease the time needed for structure repair and lower maintenance costs.

The following sections summarize the final recommendations.

6.1 For Ground model

Graphical projection of ground in 3D, together with 3D projection of an as-constructed earth structure, is in agreement with the BIM process.

- The 3D model provides an opportunity to show the range of geotechnical/ground investigation – whether this investigated zone corresponds with the volume of the ground affected by the completed structure, both in plan and in depth.
- The Geotechnical/Ground investigation report should state the reasons for the selection and range of realized tests and defend why they are the best fit for the subsequent limit state design and check.

6.2 For Geotechnical design model

The Geotechnical design model should provide the following:

- Justification that the selection of characteristic geotechnical parameters is in agreement with the potential risk (geotechnical category). For low-risk structures, the characteristic geotechnical parameters can be selected from recent experiences with similar ground, when these parameters are also part of different classification systems. For higher GCs, the characteristic values are selected from measured values, directly obtained from the lab and field tests. However, characteristic values can also be obtained from indirect field and lab tests. Such characteristic values are denoted as derived values.
- Justification that the characteristic values were selected not only for the ground (subsoil) but also for soils (or nonstandard aggregates) to be used as construction materials in the specific earth structure.
- Justification that the selection was based on a cautious estimate principle. In this case, the selection should be on the conservative side and is evaluated on the basis of number of tests performed, heterogeneity of the ground and the ability of the structure and ground to compensate for the occurrence of weaker zones with stronger ones. The verification of the sensitivity of the solved limit states to the selected values of the geotechnical parameters is the right step in this direction.

6.3 For Calculation model

- Evidence that the design approach respected potential risk (GC). For negligible risk, an up-to-date comparable experience (adoption of prescriptive measures) is sufficient. For higher risk, it is necessary to prove that for each geotechnical design situation, no relevant limit state is exceeded.
- Justification of the approach selected for limit state verification, namely the calculation model (analytical, numerical or semi-empirical), experimental model or load test or the observational method of the geotechnical design.
- Evidence that the selected calculation model is controllable and that the selected calculation model and the input data used for the model are well documented.

6.4   For Structure execution

Geotechnology plays a significant role in the construction of all types of geotechnical structures. This geotechnology should guarantee that the structure is constructed in accordance with prescribed demands. Two steps should be mentioned:

*   Structure supervision – both from the side of the investor/owner or their geotechnical expert (consultant) and from the side of the structure designer; when the main attention is focused on compaction control and soil improvement.
*   Structure monitoring – as certified that the expected structure behaviour during construction process as well during the guarantee period, respectively for the expected life span of the structure, is in the range of expected values. Monitoring can have different levels of precision and is focused on measurement of deformation (settlement), stresses and pore pressures. Results of the monitoring are also an important part of the documentation for structure authorization and statutory approval.

Finally, we stress the significance of cooperation between the individual partners, from the aspects of both engineering and sustainability. Only this complex approach can successfully fulfil the basic idea of innovation, on which modern earth structures of transport engineering are based.

References

Adam, D. & Markiewicz, R. (2001) Compaction behaviour and depth effect of the Polygon-Drum. In: Correira, A. G. & Brandl, H. (eds.) *Geotechnics for Roads, Rail Tracks and Earth Structures*. CRC Press/Balkema, Leiden, The Netherlands. pp. 27–36.

Alexiew, D. (2005) Piled embankments: Overview of methods and significant case studies. *Proc. 16th IC SMGE, Osaka*. pp. 1819–1822.

Alexiew, D., Elsing, A. & Ast, W. (2002) FEM: Analysis and dimensioning of a sinkhole overbridging system for high-speed trains at Grobers in Germany. In: Delmas, Gourc & Girard (eds.) *Geosynthetics: 7th ICG, Nice*. Swets & Zeitlinger, Lisse. pp. 1167–1172.

Atkinson, J. H. (1993) *An Introduction to the Mechanics of Soils and Foundations*. McGraw-Hill, London.

Atkinson, J. H. & Bransby, P. L. (1981) *The Mechanics of Soils: An Introduction to Critical State Soil Mechanics*. McGraw-Hill, London.

Barla, M., Atzeni, C., Pieraccini, M. & Antolini, F. (2014) Early warning monitoring of natural and engineered slopes with ground-based synthetic-aperture radar. *Rock Mechanics and Rock Engineering*, 48(1), 235–246. DOI: 10.1007/s00603-014-0554-4. Available from: https://link.springer.com/article/10.1007/s00603-014-0554-4 [accessed 8th July 2019].

Basu, D., Puppala, A. J. & Chitoori, B. (2013a) General report of TC 307 sustainability in geotechnical engineering. In: Delage, P., Desrues, J., Frank, R., Puech, A. & Schlosser, F. (eds.) *Challenges and Innovation in Geotechnics: Proc. 18th IC SMGE*, Volume 4. Presses des Ponts. Paris. pp. 3155–3162.

Basu, D., Misra, A., Puppala, A. J. & Chittori, C. S. (2013b) Sustainability in geotechnical Engineering, In: Delage, P., Desrues, J., Frank, R., Puech, A. & Schlosser, F. (eds.) *Challenges and Innovation in Geotechnics: Proc. 18th ICSMGE*, Volume 4. Presses des Ponts. Paris. pp. 3171–3174.

Beetstra, G. W. & Stoutjesdijk, T. P. (2005) *First Approach to Rational Risk Management (RRD) by the Delta Institute*, 1st November, GeoDelft, Delft.

Begemann, H. K. S. Ph. (1965) The friction jacket cone as an aid in determining the soil profile. *Proc. 6th IC SMFE, Montreal*, Volume 1. pp. 17–20.

Bishop, A. W. (1967) Progressive failure with special reference to the mechanism causing it. *Proc. EC SMFE, Oslo*, Volume 2. pp. 3–10.

Bishop, A. W. & Bjerrum, L. (1960) The relevance of the triaxial test to the solution of stability problems. *NGI Publ*. No 34, Oslo.

Bishop, A. W. & Henkel, D. J. (1957) *The Measurement of Soil Properties in the Triaxial Test*. E. Arnold, London, first edition 1957, second edition 1962.

Bizjak, K. F. & Lenart, S. (2018) Life cycle assessment of a geosynthetic-reinforced soil bridge system: A case study. *Geotextiles and Geomembranes*, 46(5), 543–558. https://doi.org/10.1016/j.geotexmem.2018.04.012

Bjerrum, L. (1967) Progressive failure in slopes of over-consolidated plastic clay and clay shales. *Journal of the Soil Mechanics and Foundations Division*, ASCE, 93(SM5), 3–49.

Boden, J. B., Irwing, M. J. & Pocock, R. G. (1979) Construction of experimental reinforced earth walls at the transport and road research laboratory. *TRRL, Supplementary Report*, 457, 162–194.

Bolton, M. D. (1979) *A Guide to Soil Mechanics*. Macmillan Press, London.

Bond, A. (2013) *Eurocode 7: Half-Term Report*. [Invited Lecture] 41st Conf. Foundation Engineering, Brno, 11th November.

Bond, A. & Harris, A. (2008) *Decoding Eurocode 7*. Taylor and Francis, London.

Brandl, H. (1999) Mixed-in-place stabilization of pavement structures with cement and additives. *Proc. 12th EC SMGE, Amsterdam, Balkema*, Volume 3. pp. 1473–1481.

Brandl, H. (2001a) The importance of optimum compaction of spoil and other granular material. In: Correira, A. G. & Brandl, H. (eds.) *Geotechnics for Roads, Rail Tracks and Earth Structures*. CRC Press/Balkema, Leiden, The Netherlands. pp. 47–68.

Brandl, H. (2001b) High embankments instead of bridges and bridge foundations in embankments. In: Correira, A. G. & Brandl, H. (eds.) *Geotechnics for Roads, Rail Tracks and Earth Structures*. CRC Press/Balkema, Leiden, The Netherlands. pp. 13–26.

Brandl, H. & Blovsky, S. (2002) Protective barriers against rockfall. *Geotechnics through Eurocode 7: Proc. 3rd Conference of the Croatian Society for Soil Mechanics and Foundation Engineering, Hvar.*

Brandl, H., Kopf, F. & Adam, D. (2005) *Continuous Compaction Control (CCC) with Differently Excited Dynamic Rollers*. Bundesministerium für Verkehr, Innovation und Technologie, Straßenforschung Heft 553, Wien. p. 150.

Briaud, J. L. (2013) *Geotechnical Engineering of Unsaturated and Saturated Soils*. Wiley, New York.

Burdo, M. C. (2016) *Final Report of Post Gradual Scholarship*. Czech Technical University, Prague.

Cazzuffi, D. & Crippa, E. (2005) Shear strength behaviour of cohesive soils reinforced with vegetation. *Proc. 16th IC SMGE, Osaka, Mill Press*, Volume 4. pp. 2493–2498.

Chamra, S. (2011) *Internal Report of the Project: Sustainable Construction*. Czech Technical University, Prague.

Chandler, R. J. & Skempton, A. W. (1974) The design of permanent cutting slopes in stiff fissured clays. *Géotechnique*, 24, 457–466.

Chang, I., Lee, M. & Cho, G. (2019) Global CO_2 emission-related geotechnical engineering hazards and the mission for sustainable geotechnical engineering. *Energies*, 12, 2567. Available from: www.mdpi.com/1996-1073/12/13/2567 [accessed 8th July 2019].

Císlerová, M., Jelínková, V., Sněhota, M. & Zumr, D. (2011) Impact of the preferential flow instability on contaminant transport in the subsurface. *Sustainable Construction, Vol. Natural Hazards*. CTU Press, Prague. pp. 179–184.

Clough, R. & Woodward, R. J. (1967) Analysis of embankment stresses and deformations. *Journal of the Soil Mechanics and Foundations Division*, ASCE, 93(SM2).

Consoli, N. C., Casagrande, M. D. T. & Coop, M. R. (2005) Behaviour of a fibre-reinforced sand under large shear strains. *Proc. 16th IC SMGE, Osaka, Mill Press*, Volume 3. pp. 1331–1334.

Correira, A. G. (2015) Geotechnical engineering for sustainable transportation infrastructure. In: *Geotechnical Engineering for Infrastructure and Development: Proc. 14th EC SMGE 2015*, Volume 1. ICE Publishing, London. pp. 49–64.

Correira, A. G., Winter, M. G. & Puppala, A. J. (2016) A review of sustainable approaches in transport infrastructure geotechnics. *Transportation Geotechnics*, 7, 21–28.

Czech Agency for Standardization (1993) *CSN 72 1002:1993: Soil Classification for Transport Structures*. Prague. (in Czech).

Czech Agency for Standardization (1987) *CSN 73 1001:1987: Ground below Spread Foundations*. Prague. (in Czech).

De Beer, E. E. (1969) Stability problems of slopes in overconsolidated clays and schists. *Proc. New Advances in Soil Mechanics I, Prague, Czechoslovak Scientific and Technical Society*. pp. 53–104.

Detert, O. & Fantini, P. (2017) High geogrid-reinforced slopes as flexible solution for problematic steep terrain: Trieben-Sunk Project, Austria. In: *Advancing Culture of Living with Landslides*, Volume 3. Springer. Cham.

Disse, M. (2015) Technical and non-technical flood defence measures in transboundary watersheds. *Proc. Int. Conf. Cooperation of Neighbouring Countries: Water Management and Flood Protection, Plasy, Czech Inst. of Civil Eng. Plasy.*

Dunlop, P. & Duncan, J. M. (1970) Development of failure in excavated slopes. *Journal of the Soil Mechanics and Foundations Division*, ASCE, 96(SM2).

Dykast, I. (1993) *Properties and Behaviour of High Clayey Spoil Heaps in the North-Bohemian Brown-Coal District.* (in Czech) PhD Thesis, Czech Technical University, Prague.

Dykast, I., Pegrimek, R., Pichler, E., Řehoř, M., Havlíček, M. & Vaníček, I. (2003) Ervěnice corridor–130 m height spoil heap from clayey material-with transport infrastructure on its surface. *Proc. 13th EC SMGE, Prague, CGtS*, Volume 4. pp. 57–76.

ELGIP Position Paper (2016) *Geotechnical Risk Reduction for Transport Infrastructure.* [Lecture] Presentation for ECCREDI, Bruxelles, 21st January.

ELGIP Position Paper (2018) *The Need for Improved Knowledge and Understanding of Ground.* Delft. Available from: www.webforum.com/elgip/web/page.aspx

EN 1744–1 (2009) *Tests for Chemical Properties of Aggregates: Chemical Analysis.* Brussels.

EN 1990–Eurocode 0 *Basis of Structural Design.* Proposal for second generation prEN 1990: 2018: Basis of structural and geotechnical design. Brussels.

EN 1997–Eurocode 7 *Geotechnical Design*, EC 7–1 and EC 7–2, CEN, (2004, 2006). Proposal for second generation prEN 1997–1: 2018. Brussels.

EN ISO 14688–1 (2002, 2018) *Geotechnical Investigation and Testing: Identification and Classification of Soil, Part 1: Identification and Description.* Brussels.

EN ISO 14688–2 (2004, 2017) *Geotechnical Investigation and Testing: Identification and Classification of Soil, Part 2: Principles for a Classification.* Brussels.

EN ISO 14689 (2017) *Geotechnical Investigation and Testing: Identification, Description and Classification of Rock.* Brussels.

EN ISO 22475–1 *Geotechnical Investigation and Testing: Sampling Methods and Groundwater Measurements.* Brussels.

Estaire, J., Pardo de Santayana, F. & Cuéllar, V. (2017) CEDEX Track Box as an experimental tool to test railway tracks at 1:1 scale. *Proc. 19th IC SMGE, Seoul.*

European Committee for Standardization (2018) *CEN/TC 396: Earthworks.* Brussels.

European Construction Technology Platform: ECTP reFINE (2012) *Research for Infrastructure Network in Europe Initiative: Building Up Infrastructure Network of a Sustainable Europe: Strategic Targets and Expected Impacts*, ECTP. Available from: www.ectp.org

Federal Highway Administration (FHWA) (2001) *Evaluating Scour at Bridges*, HEC-18, fourth edition. Report No. FHWA-NHI-01–001 (Richardson, E. V. & Davis, S. R.).

FEHRL Vision (2010) *Road Transport in Europe 2025.* Available from: www.fehrl.org/

Frank, R., Bauduin, C., Driscoll, P., Kavvadas, M., Ovesen, K. N., Orr, T. & Schuppener, B. (2011) *Designer's Guide to Eurocode 7: Geotechnical Design, EC 7–1 General Rules.* Thomas Telford, London, first edition 2007.

Frankovská, J. (2019) European standardisation in geotechnical investigation. (in Slovak) *Proc. 14th Slovak Geot. Conf.*, STU Bratislava. pp. 45–50.

Friedli, B., Hauswirth, D. & Puzrin, A. M. (2017) Lateral earth pressures in constrained landslides. *Géotechnique*, 67, 890–905.

Frydenlund, T. E. & Aaboe, R. (1994) Expanded polystyrene: A lighter way across soft ground. *Proc. 13th IC SMFE, New Delhi*, Volume 3. pp. 1287–1292.

Gazda, J. *et al.* (2008) *Commodity Exchange and Construction Waste Management.* (In Czech). CTU Press, Prague.

Girijivallanhan, C. V. & Reese, L. C. (1968) Finite element method for problems in soil mechanics. *Journal of the Soil Mechanics and Foundations Division*, ASCE, 94(SM2).

Greaves, H. (1996) An introduction to lime stabilization. *Proc. Lime Stabilization, London, Loughborough University*, Thomas Telford. pp. 5–12.

Havlíček, J. (1978) *Foundation Settlement.* (in Czech) Final report C 52–347–018, S. G. Praha.

Head, M., Lamb, M., Reid. M. & Winter, M. (2006) The use of waste materials in construction: Progress made and challenges for the future. Feature Lecture. In: Thomas, H. (ed.) *Proc. 5th ICEG*, Volume 1. Cardiff, Thomas Telford, London. pp. 70–92.

Heerten, G., Vollmert, L., Herold, A., Thomson, D. J. & Alcazar, G. (2013) Modern geotechnical construction methods for important infrastructure buildings. In: Delage, P., Desrues, J., Frank, R., Puech, A. & Schlosser, F. (eds.) *Challenges and Innovation in Geotechnics: Proc. 18th ICSMGE*, Volume 4. Paris. Presses des Ponts. pp. 3211–3214.

Hofman, R. (2019) *Geotechnics and Natural Hazards*. [Invited lecture] 14th Slovak Geotechnical Conference, STU Bratislava, 27th May.

Holeyman, A. & Mitchell, J. K. (1983) Assessment of quicklime pile behaviour. *Proc. 8th EC SMFE, Helsinki, Balkema*, Volume 2. pp. 897–902.

Holm, G., Trank, R. & Ekstrom, A. (1983) Improving lime column strength with gypsum. *Proc. 8th EC SMFE, Helsinki*, Volume 2. pp. 903–907.

Horizon 2020 Transport Advisory Group (TAG), May 2016. Available from: https://ec.europa.eu/programmes/horizon2020/en/smart-green-and-integrated-transport-%E2%80%93-work-programme-2016-2017 [accessed 8th July 2019].

ICOLD Bulletin 164 (2017) *Internal Erosion in Existing Dams, Dikes and Levees and Their Foundations*. Paris.

INTACT: European Project: On the Impact of Extreme Weather on Critical Infra Structures. (2016). Available from: https://cordis.europa.eu/project/rcn/185476/en; www.intact-wiki.eu

ISO/TR 20432:2007 *Guidelines for the Determination of the Long-Term Strength of Geosynthetics for Soil Reinforcement*. Geneva.

Janbu, N. (1973) Slope stability computations. In: *Embankment Dam Engineering*. Casagrande Volume. John Willey and Sons, New York.

Jelušič, P. & Žlender, B. (2019) Determining optimal designs for geosynthetic-reinforced soil bridge abutments. *Soft Computing*, 1–14. Available from: https://doi.org/10.1007/s00500-019-04127-8

Jirásko, D. & Vaníček, I. (2009) The interaction of groundwater with Permeable Reactive Barrier (PRB). *The Academia and Practice of Geotechnical Engineering: Proc. of 17th ISSMGE Conference, Cairo, IFOS, 2009*. pp. 2473–2478.

Jirásko, D. & Vaníček, I. (2015) Practical example of transport infrastructure protection against rock fall. In: *Geo-Environment and Construction*. Natyra, Tirana. pp. 274–283.

Jirásko, D. & Vaníček, I. (2017) Rock falls potential evaluation and protection for the road II/102 Strnady Štěchovice. In: *Building Up Efficient and Sustainable Transport Infrastructure 2017 (BESTInfra2017), Bristol: IOP Publishing Ltd*, IOP Conference Series: Materials Science and Engineering, Volume 236.

Jirásko, D., Vaníček, I. & Vaníček, M. (2017) Interaction of landslide with critical infrastructure. In: *Advancing Culture of Living with Landslides*, Volume 3. Springer. Cham. pp. 545–552.

Johnson, P. E. & Card, G. B. (1998) The use of soil nailing for the construction and repair of retaining walls. *TRL Report*, 373, 42 p.

Jones, C. J. F. P. (1996) Construction influences on the performance of reinforced soil structures. In: McGown, A., Yeo, K. C. & Andrawes, K. Z. (eds.) *Performance of Reinforced Soil Structures*. Thomas Telford, London. pp. 97–116.

Kahyaoğlu, M. R. & Vaníček, M. (2019) A numerical study of reinforced embankments supported by encased floating columns. *Acta Geotechnica Slovenica*, 16(2).

Kalsnes, B., Nadim, F. & Lacasse, S. (2010) Managing geological risk. *Proc. 11th IAEG Congress, Auckland, New Zealand*. pp. 111–126.

Karpurapu, R. & Bathurst, R. J. (1994) Finite element analysis of geotextile reinforcement retaining walls. *Proc. 12th IC SMFE, New Delhi*. pp. 1381–1384.

Katzenbach, R. (2016) *Concept of Safety and Quality Control in Geotechnical Engineering*. [Invited Lecture] Czech Geotech. Soc., Prague, 26th September.

Kempfert, H. G., Göbel, C., Alexiew, D., Heitz, C. *et al.* (1997) German recommendation for reinforced embankments on pile-similar elements. *EuroGeo 3, Munich, DGGT*. pp. 279–284.

Kézdi, A. (1964) *Bodenmechanik I., II.* Verlag für Bauwesen, Berlin.

Kjellman, W. (1948) Accelerating consolidation of fine-grained soils by means of card-board wicks. *Proc. 2nd IC SMFE, Rotterdam*. p. 302.

Kondner, R. L. (1963) Hyperbolic stress-strain response: Cohesive soils. *Journal of the Soil Mechanics and Foundations Division*, ASCE, 89(SM1).

Kopecký, M., Ondrášik, M., Frankovská, J. & Brček, M. (2019) Risk analysis of alternative trajectory of the D1 highway in the landslide territory. (in Slovak) *Proc. 14th Slovak Geot. Conf., STU Bratislava*. pp. 130–139.

Koseki, J. (2010) *Use of Geosynthetics to Improve Seismic Performance of Earth Structures*. Mercer Lecture (2010–2011). 6th IC EG, New Delhi.

Krejčí, O., Krejčí, V. & Kašperáková, D. (2019) Evaluation of the efficiency of stabilization measures on the landslides from flooding in July 1997. (in Czech) *Proc. 14th Slovak Geot. Conf., STU Bratislava*. pp. 140–149.

Kuba, P. & Wallensfeld, D. (2019) The new water line at the central wastewater treatment plant, the path from 2D design to 3D design in BIM. (in Czech) *Stavebnictví* 04/2019. pp. 54–58.

Kurka, J. & Novotná, I. (2003) The D8 motorway over old tailing dam and spoil heaps near Ústí nad Labem. In: Vaníček *et al.* (eds.) *Proc. 13th EC SMGE, Prague, CGtS*, Volume 4.

Lacasse, S. (2013) Protecting society from landslides: The role of the geotechnical engineer: 8th Terzaghi oration. *Proc. 18th IC SMGE, Paris*, Volume 1.

Lambe, T. W. & Whitman, R. V. (1969) *Soil Mechanics*. John Willey and Sons, New York.

Leflaive, E. (1985) Soil reinforcement with continuous yarns: The Texsol. *Proc. 11th IC SMFE, San Francisco, Balkema*, Volume 3. pp. 1787–1790.

L'heureux, J. S. (2019) *Landslides in Sensitive Clays*. [Lecture] ELGIP Workshop on Landslides, Paris, 2nd April.

Lind, B. B., Fallman, A. M. & Larsson, L. B. (2000) Environmental impact of ferrochrome slag in road construction. In: *Waste Materials in Construction*, Pergamon, Amsterdam.

Liška, V. (2011) *The Stock Exchange*. CTU Press, Prague.

Löfröth, H. (2019) *Methodology for Landslide Risk Mapping in a Changing Climate: Sweden*. [Lecture] ELGIP Workshop on Landslides, Paris, 2nd April.

Lunne, T., Robertson, P. K. & Powell, J. J. M. (1997) *Cone Penetration Testing in Geotechnical Practice*. Blackie Academic and Professional, London.

Malerba, P. G. (2011) Inspection and maintenance of old bridges. In: Vaníček, I. (ed.) *Sustainable Construction*. CTU press, Prague.

Mayne, P. W. (2012) Invited keynote: Geotechnical exploration in the year 2012. *Proceedings 16th Nordic Geotechnical Meeting*, Volume 1. Danish Geotechnical Society, Copenhagen. pp. 11–27.

McGown, A., Andrawes, K. Z., Hytiris, N. & Mercer, F. B. (1985) Soil strengthening using randomly distributed mesh elements. *Proc. 11th IC SMFE, San Francisco, Balkema*, Volume 3. pp. 1735–1738.

Mclnerney, J. (2016) Implementing BIM on large infrastructure projects. *Engineers Ireland Cork Region Annual Seminar 2016 BIM: Practical Examples & Learnings*. Available from: www.engineersirelandcork.ie/downloads/20162402%20EI_cork%20MM_JMcI_12final.pdf [accessed 20th April 2019].

Michalowski, R. L. & Čermák, J. (2003) Triaxial compression of sand reinforced with fibres. *Journal of Geotechnical and Geoenvironmental Engineering*, ASCE, 129, 125–136.

Morgenstern, N. R. (1969) Structural and physico-chemical effects on the properties of clays. *Proc. 7th IC SMFE, Mexico*, Volume 3. pp. 455–471.

Morgenstern, N. R. & Price, V. E. (1965) The analysis of the stability of general slip surfaces. *Géotechnique*, 15, 79–93.

Moryn, G. (2016) Geotechnical BIM: Applying BIM principles to the subsurface. *Autodesk University 2016.* Available from: http://aucache.autodesk.com/au2016/sessionsFiles/21042/12629/presentation_21042_TR21042%20Geotechnical%20BIM%20Presentation-template.pdf [accessed 20th April 2019].

Moser, A. P. (2001) *Buried Pipe Design.* 2nd edition. McGraw-Hill Professional. 544 p.

Motz, H. & Geiseler, J. (2000) Products of steel slags. In: *Waste Materials in Construction.* Pergamon, Amsterdam.

Mulligan, C. (2019) *Sustainable Engineering: Principles and Implementation.* CRC Press, Boca Raton.

Myslivec, A., Eichler, J. & Jesenák, J. (1970) *Soil Mechanics.* (in Czech and Slovak) SNTL, Praha.

O'Riordan, N. (2012) Sustainable and resilient ground engineering. Keynote Lecture, *11th Australia-New Zealand Conf. on Geomechanics, Melbourne, Australia.*

Orr, T. L. L., Matsui, K. & Day, P. (2002) Survey of geotechnical investigation methods and determination of parameter values. *Proc. Int. Workshop: Foundation Design Codes and Soil Investigation in view of International Harmonization and Performance, IWS Kamakura 2002, Tokyo, Balkema.*

Pašek, J. & Kudrna, Z. (1996) Motorway in landslide area in the České středohoří Mountains. (in Czech) *Proc. 2nd Slovak Geotechnical Conf.,* Bratislava, STU, Bratislava. pp. 97–102.

Pereira, P., Mataix-Solera, J., Úbeda, X., Rein, G. & Cerdá, A. (2019) *Fire Effects on Soil Properties.* CRC Press, Boca Raton.

Phear, A., Dew, C., Ozsoy, B., Wharmby, N. J., Judge, J. & Barley, A. D. (2005) Soil nailing: Best practice guidance. *CIRIA C637,* London. 286 p.

Pilarczyk, K. P. (1996) *Dikes and Revetments: Design, Maintenance and Safety Assessment.* Balkema, Rotterdam.

Pinard, M. I. (2001) Development in compaction technology. In: Correira, A. G. & Brandl, H. (eds.) *Geotechnics for Roads, Rail Tracks and Earth Structures.* CRC Press/Balkema, Leiden, The Netherlands.

Potts, D. M. & Zdravkovic, I. (1999, 2001) *Finite Element Analysis in Geotechnical Engineering,* Volume 1 Theory. 440 p., Volume 2 Application. 427 p. Thomas Telford, London.

prEN 1997–1:20xx: Geotechnical design. General rules. Draft April, CEN 2018. Brussels.

Proske, D. (2004) *Katalog der Risiken.* Eigenverlag, Dresden.

Raithel, M., Kuster, V. & Lindmark, A. (2004) Geotextile-encased columns: A foundation system for earth structures, illustrated by a dyke project for a works extension in Hamburg. *Proc. 14th Nordic Geotechnical Meeting.* Swedish Geotechnical Society, Ystad.

Rogbeck, Y., Lofroth, H., Rydell, B. & Andersson-Skold, Y. (2013) Tools for natural management in a changing climate. In: Delage, P., Desrues, J., Frank, R., Puech, A. & Schlosser, F. (eds.) *Challenges and Innovation in Geotechnics: Proc. 18th IC SMGE,* Volume 4. Paris.

Romancov, G. (2003) Geotechnical issues of metro construction in a historic built-up area. In: Vaníček *et al.* (eds.) *Proc. 13th EC SMGE, Prague, CGtS,* Volume 4. Prague. pp. 245–264.

Rowe, R. K. (1992) *Clayey Barriers for Mitigation of Contaminant Impact: Evaluation and Design of Barriers.* The University of Western Ontario, London, Canada.

Rowe, P. K. & Ho, S. K. (1988) Application of finite element techniques to the analysis of reinforced soil wall. In: Jarrett, P. M. & McGown, A. (eds.) *The Application of Polymeric Reinforcement in Soil Retaining Structures.* Kluwer Academic Publishers, Dordrecht. pp. 541–554.

Sarma, S. K. (1973) Stability analysis of embankment and slopes. *Géotechnique,* 23(3), 423–433.

Sarma, S. K. (1979) Stability analysis of embankment and slopes. *Journal of Geotechnical Engineering Division,* ASCE, 105(GT12), 1511–1524.

Schlosser, F. (1991) *CLOUTERRE: Soil Nailing Recommendations for Designing, Calculating and Inspecting Earth Support Systems Using Soil Nailing.* French National Research Project, ENPC, Paris. English Translation, July 1993, US Department of transportation, Federal highway administration. 302 p.

Schneider, H. (2014) Discussion. *CEN250/SC7 WG1 Meeting,* Bern.

Schofield, A. N. (1980) Cambridge geotechnical centrifuge operations. *Géotechnique,* 30(3), 225–268.

Schofield, A. N. & Wroth, C. P. (1968) *Critical State Soil Mechanics.* McGraw-Hill, London.

Sehnalová, P. (2018) *Utilization of Geothermal Energy in Engineering Structures*. Diploma Thesis. (in Czech) Czech Technical University, Prague.

Selig, E. T. (1985) Review of specifications for buried corrugated metal conduit installations. *Transportation Research Record N 1008: Culverts: Analysis of Soil-Culvert Interaction and Design*. Transport Research Board, Washington. pp. 15–21.

Sembenelli, P. & Ueshita, K. (1981) Environmental geotechnics: State of the art report. *Proc. 10th IC SMFE, Stockholm, Balkema*, Volume 4. pp. 335–394.

Sherard, J. L., Woodward, R. J., Gizienski, S. F. & Clevenger, W. A. (1963) *Earth and Earth-Rock Dams*. John Wiley and Sons, New York.

Sherard, J. L., Dunnigan, L. P. & Talbot, J. R. (1984) Basic properties of sand and gravel filters. *Journal of Geotechnical Engineering Division*, ASCE, 110(6).

Šimek, J., Jesenák, J., Eichler, J. & Vaníček, I. (1990) *Soil Mechanics*. (in Czech) SNTL, Praha.

Škara, V. (2017) *Material Characteristics of Fiber-Reinforced Fly Ash*. MSc. Thesis. (in Czech) Czech Technical University, Prague.

Skempton, A. W. & Bishop, A. W. (1954) *Soils: Chapter X of Building Materials, Their Elasticity and Inelasticity*. North Holland Publ. Co., Amsterdam. pp. 417–482.

Skempton, A. W. & Hutchinson, J. (1969) Stability of natural slopes and embankment foundations. *Proc. 7th IC SMFE, Mexico*, State of the Art Volume.

Smith, J. H. (1996) Construction of lime plus cement stabilized cohesive soils. In: Rogers, C. D. F., Glendinning, S. & Dixon, N. (eds.) *Lime Stabilization*. Thomas Telford, London. pp. 13–26.

Soga, K., Vaníček, I. & Gens, A. (2011) *Micro-Measurement and Monitoring System for Ageing Underground Infrastructures*, CTU Press (Česká technika), Prague.

Stocker, M. F. (1994) 40 years of micropiling, 20 years of soil nailing. *Proc. 13th IC SMFE, New Delhi, Oxford and IBH Publishing Co.*, Volume 5. pp. 167–168.

Stocker, M. F., Korber, G. W., Gässler, G. & Gudehus, G. (1979) Soil nailing. *Int. Conf. Soil Reinforcement, Paris, ENPC*, Volume 2. pp. 469–474.

Sutherson, S. S. (1997) *Remediation Engineering: Design Concepts*. CRC and Lewis Publishers, New York.

Svoboda, B. (2014) Impact of landslide on blasting in the quarry Dobkovičky. (in Czech) *Proc. 42nd Conf. Foundation Engineering, Brno, CGS*. pp. 30–38.

Takahashi, T. (2019) *Debris Flow: Mechanics, Prediction and Countermeasures*. CRC Press. London.

Tatsuoka, F., Tateyama, M., Uchimura, T. & Koseki, J. (1996) Geosynthetics-reinforced soil retaining walls as important permanent structures. In: De Groot, M. B., Den Hoedt, G. & Termaat, R. J. (eds.) *Geosynthetics: Application, Design and Construction*. Balkema, Rotterdam. pp. 3–46.

Taylor, N. (1995) *Geotechnical Centrifuge Technology*. Blackie Academic and Professional, Glasgow.

Terzaghi, K. (1925) *Erdbaumechanik*. F. Deuticke, Vienna.

Terzaghi, K. (1943) *Theoretical Soil Mechanics*. John Wiley and Sons, New York.

Terzaghi, K. (1959) Soil mechanics in action. *Civil Engineering*, 69, February, 33–34.

Terzaghi, K. & Peck, R. B. (1948) *Soil Mechanics in Engineering Practice*. John Wiley and Sons, New York.

Towhata, I. (2008) *Geotechnical Earthquake Engineering*. Springer-Verlag Berlin Heidelberg, Heidelberg.

Towhata, I. (2017) *Pledge by Professor I. Towhata for ISSMGE Presidency*. Available from: www.jiban.or.jp/e/wp-content/uploads/2016/12/20170620_Ikuo_Towhata_Pledges.pdf [accessed 8th July 2019].

Urbina, G. O., Egizabal, A. & San Martín, R. (2000) New trends on EAF slags management in the Basque Country. In: *Waste Materials in Construction*. Pergamon, Amsterdam.

Vaněk, P. (2011) *BIM: Not Only as a Bolt of Professions*. News and information from the Czech Chamber of Authorized Engineers and Technicians. Available from: www.zpravy-ckait.cz/4-2011/zi_1104.pdf [accessed 19th April 2019]. p. 25.

Vaníček, I. (1982) *Soil Mechanics*. (in Czech) Publishing Company of the Czech Technical University, Prague.

Vaníček, I. (1986) Behaviour of soil of the high clayey spoil heap. *Proc. 8th EC SMFE, Nurnberg, DGD*. pp. 235–240.

Vaníček, I. (1987) New ground classification composed for soils. (in Czech) *Proc. 7th Geotech. Symp., Znojmo*. pp. 19–30.

Vaníček, I. (1991) Foundation on high spoil heaps. *Proc. 10th EC SMFE, Florence*. pp. 629–632.

Vaníček, I. (1995) Ways of utilization of waste clayey material. *Proc. 10th DEC SMFE, Mamaia*. pp. 967–974.

Vaníček, I. (2004) Lessons learned from the failures of historical small dams on river-basin Lomnice. (in Czech) *Proc. Dam Days, Czech Committee of ICOLD, České Budějovice*.

Vaníček, I. (2010) Urban environmental geotechnics-construction on brownfields. Keynote Lecture. *Proc. GeoMos2010 Geotechnical Challenges in Megacities, Moscow, GRF, 2010*, Volume 1. pp. 218–235.

Vaníček, I. (2011) *Sustainable Construction*. Czech Technical University Press, Prague.

Vaníček, I. (2013a) *Present Day Position of Geotechnical Engineering*. [Lecture] Texas A & M University, College Station, 26th April.

Vaníček, I. (2013b) Civil engineering and professional prestige. (in Czech) *Stavebnictví*, No. 4 . Prague.

Vaníček, I. (2013c) The importance of tensile strength in geotechnical engineering. Šuklje Lecture. *Acta Geotechnica Slovenica*, 10(1).

Vaníček, I. (2016) *Discussion to Ground Models + Risk Evaluation and Geotechnical Categories*. [Lecture] Presentation of TG 2 on CEN250/SC7: WG 1 Meeting, Leuven, 26–27th April.

Vaníček, I. & Jirásko, D. (2017) Earth structures: Principle of sustainability. *Tecnicall*, 2. (in Czech) rektorát ČVUT.

Vaníček, I. & Škopek, P. (1989) Stability calculation of reinforced soil slope. *Proc. 12th IC SMFE, Rio de Janeiro*, 17/28. pp. 1321–1324.

Vaníček, I. & Vaníček, J. (2004) Rehabilitation of old earth dams failed during heavy floods in 2002. In: Wieland, M., Ren, Q. & Tan, S. Y. (eds.) *Proc. New Development in Dam Engineering*. Balkema. London. pp. 889–898.

Vaníček, I. & Vaníček, M. (2001) Geosynthetics reinforcement: Limit state approach. *Proc. 15th IC SMGE, Istanbul*, Volume 2. pp. 1637–1640.

Vaníček, I. & Vaníček, M. (2008) *Earth Structures in Transport, Water and Environmental Engineering*. Springer. Dordrecht.

Vaníček, I. & Vaníček, M. (2013a) Modern earth structures of transport engineering. *Procedia Engineering*, 57, 77–82. Available from: https://doi.org/10.1016/j.proeng.2013.04.012 [accessed 8th July 2019].

Vaníček, I. & Vaníček, M. (2013b) Experiences with limit state approach for design of spread foundations. In: Arnold, P., Gordon, A., Fenton, G. A. & Hicks, M. A., Schweckendiek, T. & Simpson, B. (eds.) *Modern Geotechnical Design Codes of Practice: Implementation, Application and Development*. IOS Press. Amsterdam. pp. 228–239.

Vaníček, I. & Vaníček, M. (2016) *Experiences from the High Geotextile Reinforced Retaining Wall: Case Study*. [Invited Lecture] 13th Baltic Sea Region Geotechnical Conference, Vilnius. 6 p. http://doi.org/10.3846/13bsgc.2016.039. ISSN 2424–5968, ISBN 978-609-457-956-1.

Vaníček, I., Vaníček, J. & Pecival, T. (2016) Risk evaluation of domino effect of dam failures on small river-basins. *ES Web Conferences 7, 07009: Floodrisk 2016–3rd Eur. Conf. on Flood Risk Management, Lyon*.

Vaníček, I., Jirásko, D. & Vaníček, M. (2013) Geotechnical engineering and protection of environment and sustainable development. In: Delage, P., Desrues, J., Frank, R., Puech, A. & Schlosser, F. (eds.) *Challenges and Innovation in Geotechnics: Proc. 18th IC SMGE*, Volume 4. Presses des Ponts. Paris. pp. 3259–3262.

Vaníček, I., Jirásko, D. & Vaníček, M. (2017a) Interaction of landslides with critical infrastructure: 4th World Landslide Forum, Ljubljana. *Advancing Culture of Living with Landslides*, Springer, 3, 545–551.

Vaníček, I., Jirásko, D. & Vaníček, M. (2017b) Transportation and environmental geotechnics. *Proc. Conf. TGG 2017, Saint Petersburg, Procedia Engineering*, 189, 118–125. Available from: www.sciencedirect.com/science/article/pii/S1877705817321410 [accessed 8th July 2019].

Vaníček, I., Jirásko, D. & Vaníček, M. (2018) Interaction of transport infrastructure with natural hazards (landslides, rock falls, floods). Keynote Lecture. In: Jovanovski *et al.* (eds.) *Geotechnical Hazards and Risks: Experiences and Practices, XVI DECGE*, Skopje, Volume 1. Wiley Ernst & Sohn Berlin.

Vaníček, M. (2000) Limit design approach of the reinforced soils. *Acta Polytechnica*, 40(2), 74–77.

Vaníček, M. (2003) *Contaminant Transport in the Host Rock, Numerical Modelling and Laboratory Work*. PhD Thesis, Czech Technical University, Prague. 128 p.

Vaníček, M. (2006) Risk assessment for the case of waste utilization for embankment construction. In: Logar, J., Gaberc, A. & Majes, B. (eds.) *Proc. XIII: Danube European Conference on Geotechnical Engineering*, Volume 2. Slovenian Geotechnical Society, Ljubljana. pp. 799–806.

Vaníček, M. (2017) Landslide reclamation in Šárka Valley. *Acta Polytechnica CTU Proceedings*, 10, 65–68.

Vaníček, M. (2018) Retaining wall from reinforced soil on improved subsoil. *Proc. 46th Conf. Foundation Engineering, Brno, CGtS*. (in Czech). pp. 91–96.

Vaníček, M. & Vaníček, J. (2000) Stability calculation of reinforced slopes using program SVARG. (in Czech) *Geotechnika*, 3(2), 30–31.

Van Nederveen, G. A. & Tolman, F. P. (1992) Modelling multiple views on buildings. *Automation in Construction*, 1(3), 215–224. DOI: 10.1016/0926-5805(92)90014-B

Van Staveren, M. (2006) *Uncertainty and Ground Conditions: A Risk Management Approach*. Elsevier, Amsterdam.

Vaughan, P. R. (1976) *Embankment Dams*. Teaching Scripts, Imperial College, London.

Vaughan, P. R. & Soares, H. F. (1982) Design of filters for clay cores of dams. *Journal of Geotechnical Engineering Division*, ASCE, 108(GT1).

Vodička, J., Výborný, J., Hanzlová, H. & Vytlačilová, V. (2008) Utilization of fibre-concrete in earth structures. (in Czech) *Sustainable Construction 5*, CTU Press, Prague. pp. 107–112.

Vrbová, E. (2008) *Soil Stabilization*. (in Czech) MSc Thesis, Czech Technical University, Prague. 109 p.

Winter, M., Nettleton, I. & Pritchard, O. (2019) *Innovative Geotechnical Slope Repair Techniques: Part of Highways England's Geotechnical Resilience Programme*. [Lecture] BGA and EGGS Evening Meeting, London, 6th February.

Wood, D. M. (1990) *Soil Behaviour and Critical State Soil Mechanics*. Cambridge University Press, Cambridge.

Yashima, A. (1997) Finite element analysis on earth reinforcement: Current and future. In: Ochiai, H., Yasufuku, N. & Omine, K. (eds.) *Proc. Int. Symp. Earth Reinforcement: IS Kyushu 96*, Volume 2. Balkema. Rotterdam. pp. 1111–1116.

Young, R. N., Mohamed, A. M. O. & Warkentin, B. P. (1992) Principles of contaminant transport in soils. *Development in Geotechnical Engineering*, Volume 73. Elsevier, Amsterdam.

Záruba, Q. & Mencl, V. (1969) *Landslides and Their Control*. Elsevier, Amsterdam.

Záruba, Q. & Mencl, V. (1976) *Engineering Geology*, Volume 10. Elsevier, Developments in Geotechnical Engineering, Amsterdam.

Zienkiewicz, O. C. (1967) *The Finite Element Method*. McGraw Hill, London.

Zienkiewicz, O. C. & Taylor, R. L. (2000) *The Finite Element Method*. Butterworth Heineman, Oxford.

Index

9 780367 546038